**HISTORY OF THE EARTH
AND THE ADVENT OF MAN**

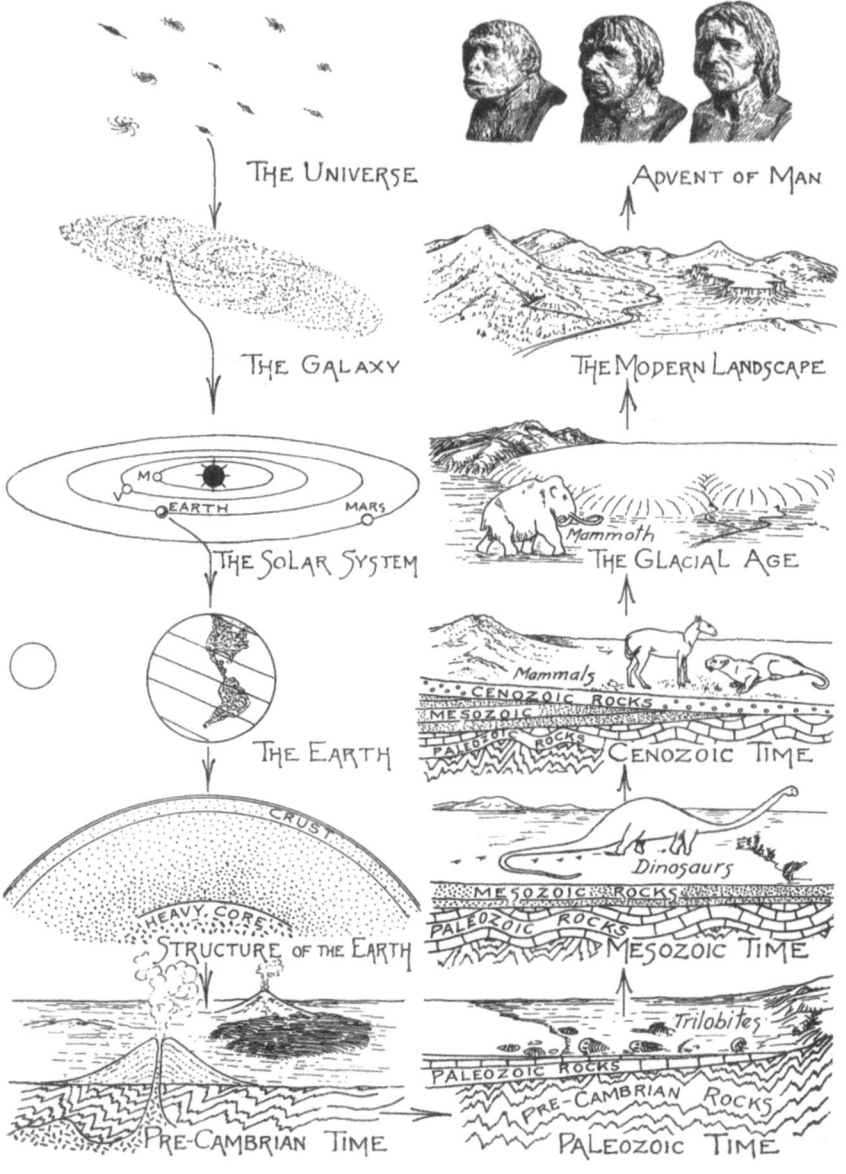

HISTORY OF THE EARTH AND THE ADVENT OF MAN

A Syllabus for Science B1, the Third Semester of a Two-Year Course in Science in Columbia College

A. K. LOBECK
RALPH L. MILLER

NEW YORK: MORNINGSIDE HEIGHTS
COLUMBIA UNIVERSITY PRESS
1937

COPYRIGHT 1937
COLUMBIA UNIVERSITY PRESS

PUBLISHED 1937

LITHOPRINTED IN THE UNITED STATES OF AMERICA

PREFACE

This Syllabus covers the course entitled Science B1, which is the first semester of the second year's work in the sciences in Columbia College.

The work of the first year in Science A1-A2, comprising physics and chemistry, is designed to establish the fundamental ideas about matter and energy. These principles are universal in time and place and have a permanent quality about them not found in the animate world.

In the second year the idea of development, change, and organization of matter becomes important. The purpose of the first semester's work, represented by Science B1, is to trace the history of the universe and of the earth from the simplest beginnings to the present. Parallel with this story is the biologic one of evolution ending with man. The subject matter of Science B1 is astronomical, geological, and anthropological. Following this, Science B2 furnishes the explanation for the life activities of plants and animals, and treats of the several fields of biology.

The combined two years of work in general science may be thought of as a drama. The first part is the preparation for the plot by the laying down of certain general facts about the inorganic world. Then comes the development of the story ending with the climax when man makes his appearance. It is probable that the two-year sequence will eventually be arranged so that the biological part will be given in the first semester of the second year with the geological and anthropological story coming at the end. This will lead naturally from the consideration of matter and energy to the living organism and finally to the evolution of the earth and man.

The laboratory activities during this semester are different from those practiced during the first year of this course when demonstrations with elaborate apparatus were possible. The study of the universe and of the earth involves concepts of such magnitude as to make it impracticable to reproduce them in the laboratory. Occasional outdoor observations of celestial objects and of geological phenomena are arranged for, however, and there is ample opportunity to work with maps, specimens, and simple apparatus. Visits will be made to the Hayden Planetarium, to the Rutherfurd Observatory, and to the exhibits in the New York museums. Students will also be encouraged to design and construct simple apparatus of their own, to observe the position of heavenly bodies, and to study such topics as the precession of the equinoxes; and they will be called upon to prepare many types of maps and diagrams to explain and to record the various kinds of climatic and geological phenomena.

Finally, it is hoped that the students will see that science must face realities and that talking and argument which may be useful in some situations will not reveal the secrets of nature. The habit of a scientific and unbiased approach to everyday problems may also be engendered.

A. K. Lobeck
R. L. Miller

Columbia University
August, 1937

LECTURE PLAN

Introduction 1
    1. Evolution.
    2. Uniformitarianism.
    3. The Scientific Method.

First Week.   THE UNIVERSE 4
    1. Size and form of the Universe. Nebulae. Galaxies.
    2. Stars. Types and Evolution of Stars.
    3. Geography of the Heavens.

Second Week.   THE SOLAR SYSTEM 8
    1. The Solar System. Origin.
    2. Members and Motions of the Solar System.
    3. The Moon. Eclipses. Tides. Precession.

Third Week.   THE EARTH 11
    1. Rotation, Revolution, Latitude, Longitude, Seasons.
    2. Time. Equation of Time. The Calendar.
    3. Climate and Weather.

Fourth Week.   EARLIEST EARTH HISTORY 15
    1. Growth of the Earth. Structure and Composition.
    2. Volcanism and Earthquakes.
    3. Minerals and Igneous Rocks.

Fifth Week.   EARLIEST ROCK-RECORDED HISTORY: PRE-CAMBRIAN . . . 19
    1. The Geologic Calendar and Its Time Units. History of the Pre-Cambrian.
    2. Sedimentation and Sedimentary Rocks.
    3. Metamorphism and Metamorphic Rocks.

Sixth Week.   AGE OF MARINE INVERTEBRATES: THE PALEOZOIC-- EARLIER STAGES 22
    1. Concepts of Historical Geology. Landmasses. Epeiric seas. Principles of Correlation.
    2. Evolution of Animals and Plants.
    3. The Early Paleozoic.

Seventh Week.   AGE OF MARINE INVERTEBRATES: THE PALEOZOIC-- LATER STAGES 25
    1. The Late Paleozoic.
    2. Life of the Paleozoic.
    3. Economic Deposits of the Paleozoic.

Eighth Week.   AGE OF REPTILES: THE MESOZOIC 29
    1. History of the Mesozoic.
    2. Life of the Mesozoic.
    3. Economic Deposits of the Mesozoic.

Ninth Week.	AGE OF MAMMALS: THE CENOZOIC	33

        1. The Cenozoic by Periods.
        2. Life of the Cenozoic.

Tenth Week.   THE GLACIAL AGE 36
        1. History of the Pleistocene (Continental Glaciation).
        2. History of the Pleistocene (Alpine Glaciation).
        3. Life of the Glacial Age. Economic Aspects.

Eleventh Week. SCULPTURING AGENTS OF THE MODERN LANDSCAPE 41
        1. Weathering.
        2. Streams.
        3. Waves, Wind, Underground Processes.

Twelfth Week. THE LARGER ELEMENTS OF THE MODERN LANDSCAPE 45
        1. Plains and Plateaus.
        2. Dome Mountains. Block Mountains.
        3. Folded Mountains. Complex Mountains. Volcanoes.

Thirteenth Week. THE PRESENT WORLD STAGE. 49
        1. North America.
        2. Europe. South America.
        3. Asia. Africa. Australia.

Fourteenth Week. ADVENT OF MAN 52
        1. Man. Origin and Early Types.
        2. Beginnings of Human Culture.
        3. Distribution of Man. Race and Culture.

Fifteenth Week. THE DISTINCTIVE ASPECTS OF HUMAN CULTURE 57

Museum Assignments and Outdoor Observations 59
        1. Hayden Planetarium.
        2. Evening Field Observations of the Sky.
        3. American Museum of Natural History: Geology, Paleontology, Mineralogy, Darwin, and Hall of Men.
        4. One-day Field Trip.

Summer Assignments. Preceding the Course 60
        1. Observation of Stars and Planets. Use of Star Map. The Astrolabe.
        2. Preparation of Original Sky Map.
        3. Special Observations Appropriate to the Year in Question.

Bibliography 61

INTRODUCTION

EVOLUTION

The keynote of this course is evolution. Matter, which was discrete at first, formed nebulae, and then stars. Stars, it is believed, pass through certain definite stages in their history, each stage endowed with certain observable physical characteristics. From the close approach of two stars the solar system had its origin and the earth came into being. Geology is concerned in part with these beginnings. Historical geology starts with the records left by the most ancient rocks. At first very fragmental, the story constantly becomes more and more nearly complete. Through two billion years it may be traced to the present. The accumulation of vast deposits of sedimentary rocks through the ages, interrupted by periods of mountain folding and continental uplift, are chapters in the story which unfolds itself through the geological periods. The whole procedure was orderly. The processes and geological activities which took place in the most remote geological past were virtually the same as those going on around us at the present time.

The sciences of astronomy and geology explain and record a series of definite and actual events and describe the results of those events. Every fact of the present has in some way been influenced by the events of the past. It is impossible fully to understand the present without knowing what has led up to it. The astronomer and geologist are considering particular and real events, events which were peculiar to themselves and happened only once; or, if they occurred more than once, did not occur under precisely the same conditions. In short, astronomy and geology must consider time and place and this makes these two sciences vastly different from physics and chemistry.

Contemporaneous with the evolution of the inorganic world was the evolution of organic forms. Evidences of life are recorded in the earliest known rocks. Life, in order to exist, must be adapted to its environment, and this has resulted in endless modification of living forms. For one or two billion years plants and animals have come down through the ages in constantly changing shapes, a process which is termed evolution. There is a vast array of evidence to show that evolution has taken place and some of the most convincing evidence is that supplied by the geologist. Evolution is accepted as a fact by all scientists. The explanation for evolution is more difficult to discover and for this there are many theories. Arguments between scientists have to do with the causes of evolution and not at all with the idea of evolution itself.

UNIFORMITARIANISM

One of the principles which dominates the natural sciences is that called uniformitarianism, as opposed to catastrophism. This means that the processes of nature, as they are going on at the present are sufficient to account for all the great changes of the past. It is only necessary that there be sufficient time. Even the oldest known geologic rocks record conditions such as now exist on the earth. Before the principle of uniformitarianism was adopted, changes in the past such as the development of canyons and the building of mountains was explained as the result of sudden catastrophes or cataclysms, and this belief prevented an understanding of what had actually occurred.

THE SCIENTIFIC METHOD, AS USED IN EARTH SCIENCES

It is impracticable in the study of astronomy and geology to obtain facts by controlled experiment after the manner of physics and chemistry. Observation of natural phenomena leads to the collection of the available facts. Then come the following steps in scientific procedure.

1. Collection of facts; their classification and arrangement.
2. Inductive reasoning from the facts, leading to the discovery of general principles or laws to explain the facts.
3. Deductive reasoning from each generalization, leading to new facts not already observed.
4. Testing of each hypothesis or law.

A. Basis of classification may be: arbitrary, that is empirical, e.g., flowers by color, minerals by hardness; or it may be explanatory, usually by taking into account the origin or genesis; hence, **genetic** classification.

B. Inductive reasoning should lead to multiple hypotheses. Detective and mystery stories introduce this technique. Intuition is often helpful, and rapid, but not always trustworthy. The inductive process is used every day by all of us, but usually not consciously and deliberately.

C. Deduction leads to new discoveries, as the discovery of Uranus, the prediction of eclipses, and the location of oil and minerals.

D. Testing of hypotheses leads to elimination of the invalid ones. Experimentation is essentially fact-finding. By controlling all of the factors, and reducing their number, actual facts can be picked out from a chaos of observations. Most valuable in physics, chemistry, and in many biological studies (such as bacteriology); but difficult in astronomy and geology because actual conditions of nature can seldom be reproduced or controlled.

Problem. Present in outline a problem from everyday life, giving the four steps leading to its solution. For instance, a man comes home and finds a window of his house broken. He formulates several hypotheses: e.g., a burglary; boys playing baseball nearby; an explosion; the awning remover; a bird flew into it, etc. Deductions and the search for further facts follow. Give an example from everyday life in which facts are confused with inferences.

THE PRESENTATION OF SCIENTIFIC MATERIAL

Discussion of the inductive and deductive methods of writing scientific papers, with an analysis of several papers or part of books to illustrate the importance of noting what method an author uses.

Problem. Select a scientific article and show how inductive and deductive methods are used in its development. Explain why readers prefer the deductive treatment.

CONTRAST BETWEEN SCIENCE A AND SCIENCE B

Science A deals with generalizations, or laws relating to matter and energy that are true at all times and in all places. Science B deals with change. Development and evolution introduce concepts of time and place.

In Science A the ideas are of general application. In Science B the ideas relate largely to particular or individual things and to particular or individual events, in the evolution of the universe and of the earth.

This idea of change and development gives the sciences of astronomy, geology, botany, zoölogy, and anthropology something in common and causes them also to adopt a technique which involves the past history of things in order to understand the present, a technique which appears unnecessary in physics and chemistry.

Problem. Show by some detailed example drawn from each of the five mentioned sciences that the history of the past is necessary in order to understand the present.

FOR STUDY AND READING

SUGGESTED
  Huxley, "Science and Education," Collected Essays, 1893-1894,
    Vol. III.
  Saidla and Gibbs, "Science and the Scientific Mind." This book contains twenty-four selections from great scientists upon scientific method, including Huxley's mentioned above, and also contains a splendid list of references.

General references on science in addition to those suggested in Science A last year:

  Planck, "Where is Science Going?"
  Wells, "The Science of Life."

FIRST WEEK

THE UNIVERSE

The earth forms the stage to which man's physical activities are confined, but his researches have been projected almost inconceivable distances through space to examine into the nature of other astronomical bodies which occupy the universe. Much is known about the nature, composition, and movements of our nearer companions on this larger stage; something is known about their origin and life history. Still less is known about our far distant companions and groups of companions so far removed that the mind has no finite terms by which their distances can be comprehended.

No play can be properly comprehended by the spectator until the action to be depicted on the stage has been given a setting in space and time. The first problem to be undertaken by the inquirer thus becomes one of orienting the world stage with respect to the larger stage of which it is a part.

A. <u>The Materials of the Universe</u>

    1. Space. What is it?
    2. Matter. The amount of matter in the known spatial universe is about equivalent to six specks of dust in Grand Central Station.
    3. Law of gravitation. Pervasive throughout the universe.

B. <u>Forms in Which Matter of the Universe Is Gathered</u>

    1. Condensations of space material to form spiral nebulae (outside our own star system) under law of gravitation.
        a. Characters of spiral nebulae and examples.
        b. Distances and size of spiral nebulae.
        c. Evolution of gaseous nebular material.
           (1) Condensation into stars and star clusters.
        d. Galaxies. Size and shape of our galaxy.
        e. Position of the earth in our galaxy explains the Milky Way.
    2. Nebulae (not spiral) within our galaxy composed of widely disseminated gaseous material.
    3. Star clouds and clusters. Star clusters are closely associated and have common motion through space.
    4. Distances of the stars. It takes light 4.16 years to reach us from the nearest star, traveling at the rate of 186,000 miles per second.

C. <u>The Stars</u>

    1. Size of the stars.
        a. The red giants, e.g., Antares with diameter 400 million miles.
        b. Stars of main sequence, comparable in size with the sun, e.g., Altair, 1.4 times size of sun.

c. Dwarf stars, e.g., Sirius B.
2. Mass of the stars.
   a. The great majority of the stars have masses from 1/5 to 5 times the mass of the sun.
   b. The red giants have extremely small densities. Antares .0000003 times the density of water.
   c. The dwarf stars have extremely high densities. Sirius B has a density 27,000 times the density of water.
   d. Stars of the main sequence have densities approximately that of water, e.g., the sun 1.4; Altair 0.6.
   e. Apparently the mass of the great majority of stars is quite constant, and where size increases, density decreases.
3. Visual magnitudes of the stars.
   a. A function of luminosity and distance.
   b. Scale of magnitude:
      (1) Stars up to sixth magnitude are visible with naked eye.
      (2) Stars to twenty-first magnitude are visible with present telescopes.
4. Double stars and eclipsing binaries.
   a. Stars which mutually revolve.
   b. Eclipsing when one passes between the earth and the other, producing diminution of light from the system.
5. Novae or new stars.
   Stars which flare up to many times their usual magnitude for short periods of time.

D. Geography of the Heavens

1. The celestial sphere.
2. Celestial poles and celestial equator.
3. Declination or celestial latitude.
   Measured north and south of the celestial equator.
4. Right ascension or celestial longitude.
   a. Measured eastward from a starting point called the vernal equinox.
   b. The meridians of celestial longitude are called hour circles.
5. Rotation of the earth on its axis causes stars to rise and set.
6. Relation of terrestrial latitude to altitude of the North Star.
7. The constellations.
   a. Naming of the constellations.
   b. Best known constellations: Ursa Major - Big Dipper, Orion, Scorpio, Leo, Southern Cross.
   c. Use of star maps and star charts to learn constellations.
   d. Use of Students' Astrolabe to locate and identify stars.

LABORATORY

The laboratory assignments for each week in this course are not designed to be entirely completed within the two-hour laboratory period. The class period is intended rather to acquaint the student with the nature of the problems and give him an opportunity to use the apparatus, models, maps, and other available material, but it will be necessary in addition for him to spend some time outside of the laboratory to put his notes into order and complete the assigned problems.

FIRST WEEK

The purpose of the laboratory the first week is to familiarize the student with the general concept of the universe, the galaxy, and the solar system, and to consider the spatial relationships of the various nebulae, stars, and planets. The sky map, the celestial globe, planet plotting chart, the "Nautical Almanac," and the Students' Astrolabe will be used to accomplish this result. The following problems are introduced as a basis from which the instructor will make individual assignments.

1. <u>Blank Sky Map</u> with list taken from the "Nautical Almanac."

   a. Plot the position of the fifty-five important stars. Label them. Use different symbols for the different magnitudes.
   <u>Note</u>: The preparation of an original sky map from direct observation of the stars, using the Students' Astrolabe, is introduced as the summer assignment preceding the course and for evening work simultaneously with this part of the course.
   b. Plot the position of the sun during the course of the year and thus determine the position of the ecliptic. Use table of positions, taken from "Almanac."
   c. Plot the position of the moon for one or two months, preferably the current month. Use table of positions taken from the "Almanac," or consult the "Almanac" itself. Determine dates of new moon and full moon and check with direct observation or with "Almanac."
   d. Plot the position of the four visible planets for one year.

2. <u>Blank Celestial Globe</u>

   a. Draw on the globe hour circles, and circles of declination.
   b. Plot from the "Almanac," or from lists provided, a dozen or so of the more prominent stars, the ecliptic, the path of the moon for a month, and of the planets for a year.
   c. Study the large celestial globe.

3. <u>Planet Plotting Chart</u>

   Study this device and do accompanying exercises, as assigned.

4. <u>Nautical Almanac</u>

   Be familiar with the nature of the contents of this book.

5. <u>The Students' Astrolabe</u>

   Be familiar with the use of this device. Set it up during the laboratory period so that it will be pointing at a certain star or planet that evening at a certain designated hour.

FOR STUDY AND READING

ESSENTIAL
   <u>Menzel</u>, "Stars and Planets," pp. 1-42, 89-106.

SUGGESTED
   Eddington, "Stars and Atoms."
   Jacoby, "Astronomy."
   Jeans, "Through Space and Time," "The Universe around Us."

SECOND WEEK

THE SOLAR SYSTEM

Concepts of the solar system have changed during the history of man's thought. These concepts have profoundly influenced his philosophy and his religion. At first the earth was believed to be the center around which all heavenly bodies moved and it has been difficult for man to accept the idea that the earth which is so important to him is one of the most insignificant bodies in space.

Only recently, too, has the idea arisen in men's minds that the earth has not always existed. Bound up with this idea were the various theories as to the origin of the whole solar system. Gradually it became clear that the various parts of the solar system were not independent units but were members of the same family with a common past.

A. Origin of the Solar System

1. Regularities of the solar system which demand explanation. These are not necessary consequences of gravitation.
    a. Orbits.
        (1) All revolve in same direction.
        (2) All are nearly in same plane.
        (3) All are nearly circular.
        (4) Regular progression of distances. Bode's Law.
        (5) Rotation of sun in same direction as revolution of planets.
        (6) Sun's equator almost parallel to planetary orbits.
    b. Planets.
        (1) The massive planets have low density.
        (2) The larger planets rotate more rapidly.
        (3) Of the four major planets the inclination of the equator to the orbit decreases with decreasing distance from the sun.
    c. Satellites.
        (1) They revolve in same direction as their planets rotate (except outer satellites of Jupiter and Saturn).
        (2) Orbits are nearly circular in the plane of the planet's equator (except as above).
2. Nebular hypothesis.
    a. Eighteenth Century. Swedenborg, Kant, and LaPlace.
    b. Slowly rotating nebulous mass cooled and contracted, rotation becoming more rapid until centrifugal force at equator threw off rings in succession, which formed the planets.
    c. Two fatal difficulties.
        (1) Rings would form many bodies.
        (2) Ninety-eight per cent of the momentum of solar system is concentrated in 1/700 of its mass, i.e., in the four major planets. Conclusion is that the angular momentum

of the planets was put into the system from the outside.
   d. Uniformitarian hypothesis gives way to catastrophic hypothesis.
3. Hypothesis of dynamic encounter.
   a. About 1900. Chamberlin and Moulton.
   b. Close approach of star to sun.
      Five to ten billion years ago.
      Tidal masses ejected. Revolved about the sun in elliptic orbits.
   c. Accounts for momentum. Accounts for sun's rotation.
   d. Two variations.
      (1) Planetesimal (Chamberlin and Moulton).
      (2) Tidal (Jeans and Jeffreys).
   e. Rotation of planets.
   f. Origin of satellites.
   g. Origin of the moon.

B. **Members and Motions of the Solar System**

   1. The Harmonic Law (Kepler).
      The squares of the periods of revolution are proportional to the cubes of the distances.
   2. Bode's Law (to explain distances).
   3. Kepler's Laws.
   4. Ptolemaic and Copernican systems.
   5. The planets considered separately.
      a. Size.
      b. Distance.
      c. Density.
      d. Satellites.
      e. Temperature and physical conditions.
   6. Planetary configurations.
   7. Comets and meteors.

C. **The Moon**

   1. Phases of the moon.
   2. Synodic month; Sidereal month; calendar month.
   3. Harvest moon; crescent moon; position of rising of full moon in winter and summer.
   4. Tides.
      a. Explanation for two high tides each day.
      b. Neap tides and spring tides.
      c. Explanation for two high tides of different heights each day.
      d. Lag of tides and other local tidal phenomena.
   5. Precession.
      a. Explanation of precession.
      b. Effect on right ascension and declination of stars.
   6. Eclipses.
      a. Solar eclipses.
         (1) Total, partial, annular.
      b. Lunar eclipses.
         (1) Total and partial.
         (2) Duration.

SECOND WEEK

LABORATORY

1. The <u>Map</u>, as a device peculiar to astronomy, geology, and geography. Relation between globe and map. The various aspects of maps and charts: projections used, scale, relief, culture, symbols. Functions of maps. Sources of maps.
    a. Map projections.
        (1) Mercator's projection.
        (2) Gnomonic or great circle chart.
        (3) Conic and Polyconic projections.
        (4) Equal area projections.

2. <u>Solar System</u>.
    This is best understood by a visit to the Copernican Planetarium at the Hayden Planetarium. The student should become familiar with the names of the planets, and general facts of the solar system.

3. <u>Phases of the Moon</u>.
    Study the tellurian.

4. <u>Tides</u>.
    Study the following Coast and Geodetic Survey publications and answer questions about tides given in the laboratory:
    "Tide Tables, Atlantic Ocean."
    "Tide and Current Tables, New York Harbor and Vicinity."
    "Tidal Current Charts, New York Harbor."

5. <u>Precession of the Equinoxes</u>.
    a. Study the instrument devised to show this phenomenon and answer questions pertaining thereto.
    b. Study the homemade instruments designed to show precession. Explain how the two forces involved produce the observed effect.
    c. Study gyroscopic tops and explain why precession appears to be sometimes in the reverse direction.

FOR STUDY AND READING

ESSENTIAL
Longwell, Knopf, and Flint, "Outlines of Historical Geology," pp. 66-73.[1]
Menzel, "Stars and Planets," pp. 43-88.

SUGGESTED
Russell, Dugan, and Stewart, "Astronomy."

---

[1] Page numbers refer to first edition; revised edition in preparation.

THIRD WEEK

THE EARTH

Most of the observable phenomena concerning the earth as a planet are due to its rotation on its axis, its revolution about the sun, and the inclination of its axis to the plane of the orbit. The various facts concerning latitude, longitude, time, seasons, and related topics, can be treated most systematically by noting how they are caused by the earth's behavior in space. These matters are directly the result of the earth's particular mode of origin and must be accounted for by any theory concerned with the origin of the solar system.

SUMMARY

A. <u>Rotation</u> (on axis). This determines:

    1. Position of axis and of equator but not the tropics nor polar circles.
    2. Oblateness, which is one cause of precession.
    3. Ferrel's Law.
    4. Latitude.
    5. Longitude.
    6. Length of the sidereal day.
    7. Celestial declination.

B. <u>Revolution</u>. This determines:

    1. Sidereal year.
    2. Celestial right ascension.

C. <u>Rotation and Revolution Together</u>. These determine:

    1. Length of the solar day.
    2. Time.

D. <u>Revolution and Inclination of Axis Together</u> determine:

    1. Position of the tropics and polar circles.
    2. Time and position of sunrise and sunset.
    3. Seasons.
    4. Climate and weather (affected by rotation also).

E. <u>Revolution and Precession of Equinoxes</u> determine:

    1. Length of tropical (or seasonal) year.
    2. Change in celestial right ascension.

A. Rotational Effects

   1. Axis. Slight variation in position of axis. Probable stability of axis during geologic time.
      Theories involving change in position of axis.
   2. Polar diameter about 7,900 miles.
      Equatorial diameter about 7927 miles.
      Oblateness the cause of precession.
   3. The Foucault pendulum experiment demonstrating earth's rotation.
   4. Ferrel's Law. Bodies in motion in the Northern Hemisphere are deflected toward the right; in the Southern Hemisphere toward the left. This is regardless of direction of movement. Most pronounced in high latitudes.
      a. Effect upon streams. Cutting right-hand bank in the case of Long Island, North Carolina, southern France, Siberia; driftwood along the Yukon.
      b. Effect on winds.
         (1) On trade winds.
         (2) On cyclones and anticyclones.
         (3) On prevailing westerlies.
      c. Effect on rifle bullets and projectiles.
      d. Effect on pendulum.
   5. Latitude.
      a. Difference in lengths of degrees.
         (1) Near equator 1° equals 68.7 miles.
         (2) Near pole 1° equals 69.4 miles.
      b. Force of gravity at different latitudes.
      c. Astronomical, geographical, and geocentric latitudes.
      d. Relation between latitude of place and altitude of polar star and altitude of sun.
      e. One meter equals 1/10,000,000 of 90° of latitude.
   6. Longitude.
      a. Difference in lengths of degrees of longitude at different latitudes.
         (1) Near equator 1° equals 69.2 miles.
         (2) At pole 1° equals 0 miles.
      b. One nautical mile equals 1' of longitude at equator, approximately equal to 1.15 statute miles.
      c. Relation between lengths of degrees of latitude and longitude on Mercator's projection.
      d. Terrestrial longitude equals the difference in time between local time and Greenwich time.
      e. Local and standard time.
      f. Time belts of the world.
      g. International date line.
   7. Sidereal day.
      a. Rising and setting of stars.
      b. Is the sidereal day longer or shorter than a solar day, and how much?
   8. Celestial declination.
      a. Effect of precession upon celestial declination.
         (1) Stars having R. A. 0 - 6 : Is the change in declination positive or negative?
         (2) Stars having R. A. 6 - 12 : Change positive or negative?
         (3) Stars having R. A. 12 - 18 : Change positive or negative?
         (4) Stars having R. A. 18 - 24 : Change positive or negative?

B. Revolution

   1. Sidereal year. How many sidereal days? How many solar days?
   2. Celestial right ascension.
      Apparent eastward movement of sun among the stars.

C. Rotation and Revolution

   1. Solar day. Is it longer or shorter than a sidereal day and how much? Mean solar day.
   2. Time.
      a. Equation of time. Causes.
         (1) Effect of eccentricity of orbit.
         (2) Effect of obliquity of ecliptic.
      b. Sundials.

D. Revolution and Inclination of Axis

   1. Tropics and polar circles.
      a. Altitude of sun at different seasons.
      b. Area of earth's surface included in
         (1) Tropical belt.
         (2) Intermediate belt.
         (3) Polar zones.
      c. Relative amount of insolation received at equator and poles at different seasons.
   2. Time and position of sunrise and sunset.
      a. At different seasons and at different places.
      b. Apparent path of the sun in the sky at different latitudes at different seasons.
      c. Midnight sun.
   3. Seasons.
      a. March equinox.
         (1) Areas of daylight and darkness.
         (2) Relative length of daylight and darkness.
         (3) Apparent path of the sun.
      b. June solstice.
      c. September equinox.
      d. December solstice.
   4. Climate and weather.
      a. Wind belts.
         (1) Equatorial. Climatic elements; conditions of
            (a) Pressure.
            (b) Wind.
            (c) Precipitation.
            (d) Temperature.
         (2) Trade winds, north and south.
         (3) Tropical calms, north and south.
         (4) Prevailing westerlies.
         (5) Polar lows.
      b. Climatic types.
         (1) Low latitude.
            (a) Tropical rain forest.
            (b) Savanna.
            (c) Desert.

                (2) Middle latitude.
                    (a) Mediterranean.
                    (b) Wet subtropical.
                    (c) Marine west coast.
                    (d) Humid and dry continental types.
                    (d) Sub-Arctic.
                (3) High latitude. Polar climate.
            c. Weather.
                (1) Cyclones and anticylcones.
                (2) Thunderstorms.
                (3) Tornadoes.
                (4) Hurricanes and typhoons.

E. Revolution and Precession

    1. Tropical or seasonal year. Is it longer or shorter than sidereal
       year and how much?
    2. Change in right ascension. Is this a positive or negative change,
       and are all stars affected the same way?
    3. Effect of precession upon climate.

                              LABORATORY

        By means of various pieces of apparatus, maps, and diagrams
the student will be expected to grapple with the essential elements
of the subject and acquaint himself with the fundamental ideas.
        Demonstrations explaining latitude, longitude, sidereal and
solar days and years, equation of time, Foucault's rendulum, Ferrel's
Law will be arranged.
        One hour of the period will be devoted to an exercise on cli-
mates of the world. Certain phenomena and observations will be pre-
sented from which the student will be expected to classify the climates
in question.

                         FOR STUDY AND READING

ESSENTIAL
    Earth in Space. Assignments to be indicated.
    Climates of the World. Assignments to be indicated.
    Equation of Time Exercise Sheet and Slide Rule. Assignments to be
        indicated.
    Sun Altitude Indicator. Assignment to be indicated.

SUGGESTED
    Finch and Trewartha, "Elements of Geography."
    Kendrew, "Climates of the World."
    Russell, Dugan, and Stewart, "Astronomy."

FOURTH WEEK

EARLIEST EARTH HISTORY

In human history the earliest events are obscure because of the lack of documentary evidence. Similarly in earth history, the first pages in the story are so far removed in time (two billion years or more) that rock-recorded evidence is no longer available. The original crust of the earth is nowhere exposed to study. If an original crust existed it is now everywhere destroyed or concealed by later rock deposits. Consequently these early stages in the evolutionary story of the earth must be approached by scientific speculation, carefully controlled by the known facts on the behavior of astronomic bodies and of the physical constituents of which the earth is composed. After these first speculative stages, the story becomes rapidly clearer, so that the geologist can tell what the earth looked like 400 million years ago nearly as accurately as the historian can describe the appearance of Rome in Caesar Augustus' reign.

Among the earliest phenomena which operated to alter the original aspect of the planet are to be listed volcanic activity, earthquakes, and the fusion and solidification of crustal material to produce igneous rocks. As these activities still continue we have a means of observing processes which have been going on since the remote past.

A. Development and Composition of Crust

1. Planetesimal hypothesis.
    Growth of earth as a solid body by infall of planetesimals.
2. Gaseo-tidal hypothesis.
    Solidification of earth from gaseous state.
3. Composition and density zoning of earth's interior.
    a. Evidenced by
        (1) Calculation of weight of earth.
        (2) Speed of transmission of earthquake waves.
        (3) Composition of meteorites.
    b. Section through earth showing density and composition of different zones.
4. Birth of hydrosphere and atmosphere.
    a. Under planetesimal hypothesis.
    b. Under gaseo-tidal hypothesis.
5. Isostasy or equilibrium of the earth's crust. Its possible rôle in development of continents and ocean basins.
6. Composition of crust of the earth by weight.
    a. Rocky crust 93 per cent of total.
    b. Hydrosphere 7 per cent of total.
    c. Atmosphere .03 per cent of total.
7. Composition of crust by elements.
    Oxygen 50.02 per cent
    Silicon 25.80 per cent
    Aluminium 7.30 per cent.

Iron 4.18 per cent.
Calcium 3.22 per cent.
Magnesium 2.08 per cent.
Sodium 2.36 per cent.
Potassium 2.28 per cent.
All others 2.16 per cent.

B. Volcanism

1. Source of heat to produce volcanism.
   a. Original heat.
   b. Heat from radioactivity.
   c. Heat produced by friction during earth movements.
2. Phenomena of volcanism.
   a. Composition of magmas.
      (1) Parent magma of subcrust basic in composition.
      (2) Acidic magmas produced by
         (a) Gravitative differentiation.
         (b) Liquid immiscibility.
         (c) Assimilation.
   b. Magmas work toward surface by
      (1) Abyssal injection.
      (2) Stoping.
3. Types of volcanic eruptions.
   a. Explosive.
      (1) Characteristics of explosive eruptions.
      (2) Cinder cone.
      (3) Ejectamenta. Gases, bombs, lapilli, ash, pumice, scoria.
   b. Intermediate.
      Characteristics of intermediate type. Combination of explosive characters and quiet characters.
   c. Quiet.
      (1) Characteristics of quiet eruptions.
      (2) Shield cones.
      (3) Lavas. Pahoehoe and aa.
   d. Fissure eruptions.
4. Volcanic belts of the world.

C. Earthquakes

1. Origin.
   a. Strength of crust.
   b. Accumulation of stresses.
   c. Elastic rebound hypothesis.
   d. Faults. Tension or normal faults, and compression or reverse and thrust faults.
2. Phenomena of earthquakes.
   a. Preliminary and long waves.
   b. Primary and secondary waves.
   c. Transmission of earthquakes demonstrates earth zoning.
   d. Seismographs and seismograms.
   e. Location of epicenter of earthquake.
   f. Rossi-Forel scale of earthquake intensity.
   g. Frequency of earthquakes.
   h. Seismic belts of the world.

FOURTH WEEK

D. **Minerals and Igneous Rocks**

1. Distinction between rocks and minerals.
2. Properties of minerals.
   a. Chemical composition.
   b. Crystal form.
   c. Color, streak, luster.
   d. Hardness. Scale 1-10.
   e. Cleavage.
   f. Specific gravity.
3. General chemical classification.
   a. Native elements; oxides; sulfides; silicates; carbonates; sulfates; and other minor types.
   b. Frequently minerals are grouped into metallics and non-metallics.
4. Common rock-forming minerals to be learned. Quartz, feldspar (orthoclase and plagioclase), mica, hornblende, pyroxene, olivine, calcite, dolomite.

E. **Igneous Rocks**

1. Classification of rocks. Igneous, sedimentary, metamorphic.
2. Basis of classification of igneous rocks--texture and composition.
   a. Glassy rocks. Obsidian and pumice.
   b. Felsitic rocks. Acidic (rhyolite) through intermediate (andesite) to basic (basalt).
   c. Porphyritic rocks. Contain large phenocrysts.
      (1) Granite porphyry.
      (2) Diorite porphyry.
   d. Coarse-grained, even, granular rocks ranging from acidic (granite) through intermediate (syenite, diorite) to basic (diabase, gabbro).
3. Occurrence of igneous rocks.
   a. Intrusive (usually coarse grained). Batholiths, dikes, sills, sheets, laccoliths, stocks, volcanic necks.
   b. Extrusive (usually fine grained to glassy). Lava flows, tuffs, breccias.

LABORATORY

Mineralogy

For everyday geologic work the ability to recognize the common minerals at sight or by the application of simple tests is of great value. These tests depend on the physical and chemical properties of minerals, a number of which will be examined in the laboratory. About twenty-five common minerals are supplied in the laboratory set. These will be tested by the student for some of the following properties, and the minerals will be identified thereby:

1. <u>Hardness</u>. The hardness of minerals is expressed in terms of a scale from 1 to 10. A mineral will scratch every other mineral lower in the scale and can be scratched by every mineral higher in the scale.

2. **Cleavage**, or the property of splitting along closely spaced parallel planes. Minerals may cleave in as many as three directions.
3. **Color and Streak.** The apparent color of a mineral is seen in the specimen itself; the true color by obtaining its streak on a porcelain plate.
4. **Specific gravity.** The simplest way to test the specific gravity is by hefting the specimen in one's hand.
5. **Magnetism.** Ability of some minerals to attract a magnet.

FOR STUDY AND READING

ESSENTIAL
Longwell, Knopf, and Flint, "Outlines of Historical Geology," pp. 66-73; "Outlines of Physical Geology," 168-86, 187-220, 247-62, 296-99, 334-43.

SUGGESTED
Agar, Flint, and Longwell, "Geology from Original Sources," Ch. I, IX, and X.
Chamberlin, "The Origin of the Earth."
Heck, "Earthquakes."
Newman, "The Nature of the World and Man," Ch. II.

FIFTH WEEK

EARLIEST ROCK--RECORDED HISTORY: PRE-CAMBRIAN

With the first formation of rocks which have survived through geologic time, and are still available for study and interpretation, historical geology really begins. The record of the rocks forms the material from which the geologic events are deciphered. Major events in the story have been used as time markers from which a geologic calendar has been built, in much the same way that the barbarian invasion of the Mediterranean countries is used by historians in separating the classical period of ancient history from the Dark Ages of medieval history. The geologist is now able to date events in terms of years quite accurately, but he finds it more informative to refer to Pre-Cambrian time, or the Paleozoic era, much as the historian refers to ancient history or medieval history instead of saying before or after 400 A.D.

With the beginning of the rock record, the earliest knowledge of important geological activities becomes available. Chief among these is the deposition of material in inland seas to form sedimentary rocks, and the alteration of sedimentary and igneous rocks by pressure to form metamorphic rocks.

A. <u>The Geologic Calendar and Its Time Units</u>

    1. Methods of measuring geologic time.
       a. Salinity of the ocean.
       b. Rate of deposition of sediments.
       c. Counting of varves.
       d. Rate of erosion.
       e. Breakdown of radioactive elements.
    2. Length of geologic time in years.
    3. Subdivisions of geologic time.
       a. Eras. Delimited by world-wide mountain-making revolutions.
       b. Periods. Delimited by widespread retreat of seas from continents and local mountain-making disturbances.
       c. Epochs. Delimited by minor retreats of seas.
    4. Names on the geologic calendar.
       Learn the calendar on pages 8-9 of "Outlines of Historical Geology."[1]

B. <u>History of the Pre-Cambrian</u>
    1. Pre-Cambrian time covers two-thirds of the geologic calendar.
    2. Archeozoic era.
       a. Thick sediments deposited in Great Lakes region (Coutchiching, Grenville, Keewatin).
       b. Granites (Laurentian) intruded.
       c. Temiskaming sediments.
       d. Algoman revolution. Intrusion of gold deposits in Ontario.

---

[1] Page numbers refer to first edition; revised edition in preparation.

3. Proterozoic era.
    a. Huronian sediments containing iron deposits of Lake Superior region.
    b. Keweenawan sediments and volcanics, and Michigan copper deposits.
    c. Killarnean revolution and Lipalian erosion interval.
4. Life of the Pre-Cambrian.
    a. Direct evidences.
        (1) "Eozoon canadense."
        (2) Calcareous algae.
        (3) Radiolaria and sponges.
    b. Indirect evidences.
        (1) Abundant limestone.
        (2) Graphite in sediments.
        (3) High order of development of earliest Cambrian fossils.

C. <u>Sedimentation and Sedimentary Rocks</u>

1. Definition of sediment.
2. Agents which transport and deposit sediments.
    a. Water. Streams, waves, currents.
    b. Wind.
    c. Ice.
3. Examples of sediments being deposited now.
    a. Marine sediments.
        (1) Beach gravels and sands.
        (2) Continental shelf sands and muds.
        (3) Deep sea muds and oozes.
        (4) Coral reefs and coquinas.
    b. Continental deposits.
        (1) Delta and flood plain deposits.
        (2) Alluvial fans and alluvial plains.
        (3) Wind and ice-deposited sediments.
        (4) Salt deposits in drying lakes and enclosed seas.
        (5) Organic deposits.
4. Consolidation of sediments to form:
    a. Conglomerate, sandstone, shale, limestone, and other types of sedimentary rocks.
    b. Chemical composition of these sediments.

D. <u>Metamorphism and Metamorphic Rocks</u>

1. Definition of metamorphism.
2. Agents of metamorphism.
    a. Heat.
    b. Pressure.
    c. Gaseous and liquid emanations from magmas.
3. Types of metamorphism.
    a. Contact metamorphism.
    b. Dynamic metamorphism.
    c. Load metamorphism.
4. Characteristics of metamorphic rocks.
    a. Recrystallization.
    b. Foliation and cleavage.
    c. Cementation and induration.

5. Types of metamorphic rocks.
    a. Slate, phyllite, and schist derived from shales.
    b. Quartzite from sandstone.
    c. Marble from limestone.
    d. Gneiss from granite.

LABORATORY

Rocks

The common types of rocks are recognizable at sight. They carry telltale marks which indicate whether they are igneous, sedimentary, or metamorphic. Within each of these three divisions schemes of classification have been developed which enable the student to identify each of the specimens in the rock set by its correct name.

In addition to classification, many rocks show structures which indicate the mode and environmental conditions of its origin. The laboratory exercises for this week's work include the following types of exercises:

1. Classification of igneous rocks according to mineral content, texture, and dominant color.

2. Large-scale igneous rock structures and relative ages of igneous rocks, as worked out from the principle that an igneous rock is younger than everything it intrudes.

3. Environment of igneous rocks.

4. Classification of sedimentary rocks according to grain size and origin.

5. Sedimentary rock structures which indicate environment in which the rock was formed.

6. Metamorphic rock classification determined by structure, texture, and composition. Progressive metamorphism showing degrees of intensity of metamorphism.

FOR STUDY AND READING

ESSENTIAL
  Longwell, Knopf, and Flint, "Outlines of Historical Geology,"
            pp. 6-15, 74-100.[1]
        "Outlines of Physical Geology,"
            pp. 144-67, 263-77.

SUGGESTED
  Agar, Flint, and Longwell, "Geology from Original Sources," Ch. VIII and XI.
  Grabau, "Historical Geology," Ch. XXX.

---

[1] Page numbers refer to first edition; revised edition in preparation.

SIXTH WEEK

AGE OF MARINE INVERTEBRATES: THE PALEOZOIC--EARLIER STAGES

With the beginning of the Paleozoic era the story becomes much clearer. The geography of lands and seas can be worked out, sources of sediments accurately determined, and even climates can be inferred. Much of this knowledge depends on the recognition of the age of deposits, and the correlation of formations in widely separated regions.

Also with the opening of the Paleozoic era, the fossilized remains of former life become abundant and the clear record of earlier inhabitants of the earth is well preserved. Study of this record reveals an orderly development of higher forms of life from lower ones, according to the principles of organic evolution, first clearly enunciated by Darwin.

A. <u>Concepts of Historical Geology</u>

    1. Geanticlines (land masses) and geosynclines.
       a. Geanticlines are rising areas, supplying sediments.
       b. Geosynclines are sinking areas receiving sediments.
       c. Isostasy with respect to movement of geanticlines and geosynclines.
       d. Neutral continental areas.
    2. Epeiric seas.
       Modern examples. Baltic Sea, Hudson Bay.
    3. Law of superposition.
    4. Law of unconformability.
    5. Correlation by:
       a. Continuity of formations.
       b. Relative succession of beds.
       c. Disconformities and unconformities.
       d. Lithologic similarity.
       e. Fossils.
          (1) What fossils are.
          (2) Method of preservation.
          (3) Types.
          (4) Uses in correlation.

B. <u>Evolution</u>

    The principle of organic evolution is accepted as a fact by all scientists. Evidence comes from many diverse fields of study. However, the explanation of the principle furnishes a problem which is the subject of controversy, and upon which much work is still being done.

    1. Evidences of evolution.
       a. Geologic and paleontologic evidence. Development of complex

organisms from earlier simple forms traceable in fossil
records of the rocks, e.g., evolution of the horse.
- b. Geographic distribution. Isolation as a factor in differentiation.
- c. Classification. Gradations between species. Subspecies and varieties.
- d. Artificial selection. Breeding of new species. Examples.
- e. Structure of animals and plants. Comparative anatomy. Relationships between groups of animals and plants shown by homologous structures. Primitive organs. Vestigial organs. Blood tests.
- f. Embryology. Racial development revealed in development of embryos of individuals. Usually expressed: "Ontogeny recapitulates Phylogeny."

2. Explanations of evolution.
- a. Lamarckism. Structures developed through use and lost by disuse. Inheritance of acquired characteristics involved.
- b. Darwinism. Natural selection of chance variations which are useful in survival.
- c. Mutations (de Vries).
- d. Mendelian Law. Dominant and recessive characteristics.
- e. Orthogenesis.

3. Classification of plants and animals.

C. Paleozoic Era by Periods

1. Cambrian (from Cambria in Wales).
   - a. Seas invaded and covered Appalachian trough and Cordilleran trough in Lower and Middle Cambrian time, depositing thick sediments.
   - b. In Upper Cambrian time seas spread over interior of country depositing limestone and dolomite.
   - c. Mountain building, of local importance only, occurred in Vermont at the end of Cambrian. Called "Green Mountain disturbance." Seas retreated, except from Appalachian trough.

2. Ordovician (from Ordovices, a Welsh tribe).
   - a. Seas began spreading widely in Ordovician, reaching their maximum extent in Middle Ordovician time (Trenton time). Deposits mainly limestone.
   - b. In Upper Ordovician (Richmond time) another widespread sea. Sediments shales and sandstones near the land masses (Appalachia and Cordillera); limestones in the interior.
   - c. Ordovician closed with complete retreat of seas and intense but localized mountain building in Quebec and New England, called the Taconic Disturbance.

3. Silurian (from Silures, a Welsh tribe).
   - a. Coarse conglomerates, sandstones, and red shales near Appalachia (raised by Taconic Revolution) give way inland to limestone deposition, where clear widespread seas again prevail (Niagaran limestone).
   - b. In late Silurian arid conditions prevail in east with important salt deposits and red beds in New York State.
   - c. No mountain building in North America at close of Silurian, but Caledonian Revolution in other continents.

SIXTH WEEK

LABORATORY

Zoölogical Classification and Evolution

The usual classification schemes of plants and animals appear to be composed of arbitrary and unrelated subdivisions into which species are pigeonholed. This is because space limitations usually do not permit the expression of genetic relationships. Construction of a life-tree, however, shows not only how different forms are related to each other, but also it stresses the evolutionary development through geologic time. The laboratory work this week will consist of some of the following exercises:

1. Linnaean scheme of biologic nomenclature with examples of its subdivisions. Construction of an evolutionary tree for the animal kingdom and location of common types of fossils and of present-day animals on that tree. Emphasis will be placed on the line of descent of man.

2. Study of suites of specimens of a similar evolutionary tree for the plant kingdom.

3. Study of suites of specimens and of pictures showing stages in the evolutionary development of such forms as the ammonites, trilobites, cephalopods, the elephant, the horse, the rhinoceros, camel.

4. Construction of charts showing the rise, dominance, and decline of important groups of animals through geologic time.

FOR STUDY AND READING

ESSENTIAL
  Longwell, Knopf, and Flint, "Outlines of Historical Geology,"
    pp. 16-65, 101-18, 119-56.[1]

SUGGESTED
  Darwin, "Origin of Species."
  Grabau, "Historical Geology," Ch. XXXI-XXXIV.
  Mason, "Creative Evolution."
  Newman and others, "The Nature of the World and Man," Ch. XIII.

---

[1]Page numbers refer to first edition; revised edition in preparation.

SEVENTH WEEK

AGE OF MARINE INVERTEBRATES: THE PALEOZOIC--LATER STAGES

The geologic events, which occurred in the latter part of the Paleozoic, form a pattern which in its larger aspects is quite similar to that of the earlier Paleozoic. Hence the two-cycle idea. Life forms, however, showed continued advance, with land plants and land animals first appearing and spreading rapidly. The most important coal deposits of the world were formed during this time. Since coal is by far the most valuable mineral product to man, special study is made of its origin and occurrence.

A. Paleozoic Era by Periods (Continued)

 1. Devonian (from Devonshire, England).
     a. Limestone seas in early Devonian become increasingly muddy as Appalachia is uplifted in later Devonian.
     b. Catskill delta built out from Appalachia in late Devonian.
     c. Acadian Disturbance in southeastern Canada and northern New England.
 2. Mississippian (from Mississippi Valley).
     a. Widespread limestones in central and western United States.
     b. Continental sandstones and red shales in east.
     c. Extensive limestones in far west (Red Wall and Madison).
     d. Complete retreat of seas at close of Mississippian. Uplift of Appalachia, Llanoria, and Siouxia.
 3. Pennsylvanian (from Pennsylvania).
     a. Thick continental sandstones deposited over eastern United States. Occassionally extensive swamps prevailed in which plant remains were preserved to form coal deposits.
     b. Extensive limestones in inland seas of western United States.
     c. Ouachita Disturbance in Arkansas and Oklahoma during the Pennsylvanian; Marathon Disturbance in Texas at close of the period.
 4. Permian (from Perm, Russia).
     a. Continental conditions in eastern United States.
     b. Red beds and gypsum deposits, the last stage of evaporating landlocked seas in southwestern United States.
     c. Limestone deposits in Cordilleran trough.
     d. Widespread glaciation in Permian.
     e. Permian closed by widespread emergence of continents and mountain building in many parts of the world; called Appalachian Revolution.

B. Résumé of the Paleozoic
   The Paleozoic era may be divided into two cycles with essentially similar histories. The first cycle is composed of the Cambrian, Ordovician, and Silurian periods; the second of the Devonian, Mississippian, Pennsylvanian, and Permian. In each case the cycle starts with quiescent conditions and widespread limestone

seas (Cambrian-Ordovician in first cycle; Devonian in second), followed by more restricted seas, continental conditions of sedimentation, glaciation, increased aridity (salt and red beds of Silurian and Permian), and closed by mountain building of worldwide extent (Caledonian Revolution and Appalachian Revolution).

The student should fit into this scheme other events of similar nature in the two cycles. Careful study of the paleogeographic maps in the text is recommended.

C. Life of the Paleozoic

   1. Cambrian.
      a. Trilobites dominated Cambrian seas and showed a high degree of development.
      b. Brachiopods were abundant but usually of simple types.
      c. Sponges and calcareous algae, gastropods.
   2. Ordovician.
      a. Cambrian forms persist. Trilobites still abundant but relatively less important.
      b. New forms. Cephalopods; the primitive armored fishes; graptolites most abundant in this period.
   3. Silurian.
      a. Continuation of Ordovician forms.
      b. Scorpions and eurypterids, first air breathers.
   4. Devonian.
      a. Trilobites begin to show old-age characteristics.
      b. Brachiopods very numerous.
      c. Fishes abundant; had interior skeletons and paired fins. Crossopterygians and Dipnoi point to first terrestrial vertebrates.
      d. First record of Amphibia.
      e. First large forests.
   5. Mississippian.
      a. Brachiopods, crinoids very abundant.
      b. Sharks rulers of seas.
      c. Trilobites definitely on the wane.
   6. Pennsylvanian.
      a. Amphibians abundant.
      b. First reptiles.
      c. Great forests of treelike ferns.
      d. Many enormous insects.
   7. Permian.
      a. Culmination of the Amphibia.
      b. Primitive reptiles, some of which show mammalian characters.
      c. Glossopteris type of flora (cold weather).

At the end of the Permian most of the archaic forms of life became extinct. Trilobites died out completely. Many of the other groups entered the Mesozoic much diminished in numbers, only the less specialized forms being able to withstand the world-wide effects of the Appalachian Revolution.

D. **Economic Deposits of the Paleozoic**

 1. Coal.
    a. Origin.
    b. Grades and ranks determined by content of fixed carbon.
       Peat, lignite, bituminous coal, anthracite, graphite.
    c. Coal regions of the world.
    d. Methods of mining coal.
 2. Building stones.
    a. Marble.
       (1) Origin.
       (2) Productive localities.
       (3) Quarry methods
    b. Limestone.
       (1) Origin and utilization.
       (2) Indiana limestone.
    c. Other building stones.
 3. Silurian salt deposits.
 4. Metallic ore deposits.
    a. South African gold deposits.
    b. Clinton iron ore deposits.
    c. Tri-state lead and zinc deposits.

LABORATORY

Fossils

"A fossil is the remains or traces of a plant or animal preserved in the rocks of the earth." The mode of preservation of fossils is considered first in the laboratory, with illustrative specimens of replaced and permineralized remains, and preservation by distillation of volatile matter. The distinction is made between fossils which are internal molds, external molds, and casts of the original.

Study is made of typical specimens of fossils from the plant kingdom and of invertebrate animals. Pictures are used to illustrate fossils of the vertebrate animals. Emphasis will be placed on the utilization of fossils in geologic work, that is, for the determination of ages of sedimentary rocks.

The science of paleontology involves many kinds of techniques and is highly specialized.

Micropaleontology is the study of small forms hardly visible to the naked eye. These are the foraminifera, ostracodes, and minute crustacea which are so important to the identification of beds or "horizons" in the oil fields.

Invertebrate paleontology is the study of the vast number of the smaller organisms which are preserved in rocks of all geologic ages. Some of these, notably in the Cambrian, are the remains of jelly-fish and sea worms, preserved only as lacy films of carbon.

The study of brachiopods, molluscs, corals, crustacea, and similar intermediate-sized forms occupies many paleontologists who are interested in numerous minute differences and gradations which help to explain the evolutionary changes from one geologic period to another.

Vertebrate paleontology is carried on only by large museums able to send great expeditions into the field. The collecting and preparing of specimens is very expensive. Important only in connection with the Mesozoic and Cenozoic eras.

Paleobotany finds its value especially in the study of Cretaceous and Cenozoic floras. Some plant remains are marvelously well preserved and throw much light upon the evolution of plants.

FOR STUDY AND READING

ESSENTIAL
   Longwell, Knopf, and Flint, "Outlines of Historical Geology,"
      pp. 157-89;[1] "Outlines of Physical Geology," pp. 311-14.

SUGGESTED
   Grabau, "Historical Geology," Ch. XXXV-XL.
   Newman and others, "The Nature of the World and Man," Ch. VIII,
      IX, X.

---

[1] Page references refer to first edition; revised edition in preparation.

EIGHTH WEEK

AGE OF REPTILES: THE MESOZOIC

The Mesozoic era is a time of transition between the ancient geography and life of the Paleozoic which seems quite unrelated to the world of today, and the more modern-looking continents and life forms of the Cenozoic era. The Mesozoic has commonly been called the Age of Reptiles because of the size and abundance of this class of animals, especially of the dinosaurs who conquered and dominated in all three media of land, air, and water. More important to man, however, was the formation in Mesozoic time of rich deposits of petroleum. The iron deposits of the Pre-Cambrian, the coal deposits of the Paleozoic, and the oil deposits of the Mesozoic and Cenozoic are to a considerable extent responsible for the high degree of industrialization of the United States.

A. **History of the Mesozoic Era**

 1. Triassic (from Germany, meaning three divisions).
    a. North America largely emergent during Triassic time.
    b. Great thicknesses of sandstones and shales deposited in landlocked basins of eastern United States.
    c. Desert deposits in southwestern United States.
    d. Marine Triassic deposits confined to Cordilleran geosyncline.
    e. Palisades Disturbance in late Triassic time, manifested by igneous intrusions and extrusions in eastern United States.
 2. Jurassic (from Jura Mountains, France and Switzerland).
    a. No Jurassic deposits east of the Mississippi. Jurassic was a period of profound erosion in this area.
    b. Marine invasions of the Cordilleran and Rocky Mountain region in Upper Jurassic.
    c. Desert deposits in Arizona, Utah, and Colorado.
    d. Mountains formed along Pacific coast and extensive batholiths intruded in California and Idaho at close of Jurassic (Nevadian Revolution).
 3. Cretaceous (from Creta meaning chalk).
    a. Lower Cretaceous (Comanchean).
       (1) No deposits in eastern United States.
       (2) Marine invasions from the Arctic and Gulf of Mexico covered Great Plains and Rocky Mountain region with limestone and chalk deposits.
       (3) West coast almost continuously flooded during Lower Cretaceous.
    b. Upper Cretaceous (Gulfian).
       (1) Upper Cretaceous seas the greatest in North America since the Devonian.
       (2) Deposits for the first time along the present Atlantic Coastal Plain (sandstones and marls).
       (3) Gulf Coastal Plain continuously submerged.

(4) Extensive marine invasion of Great Plains and Rocky Mountain states connecting with both Arctic Ocean and Gulf of Mexico.
   (5) Laramide Revolution at the end of the Cretaceous closed the Mesozoic era. World-wide mountain building, manifested in North America by the rise of the Rocky Mountains and renewed uplift of the Appalachian region.
4. Resumé of the Mesozoic.
   The Mesozoic era contrasts with the two cycles of the Paleozoic, when quiescent conditions and widespread seas prevailed first. The Mesozoic era followed directly upon a time of pronounced uplift and mountain building (Appalachian Revolution), so that the opening chapters (Triassic and Jurassic) are periods of widespread continental conditions, with fluvial, lacustrine, and desert deposits dominant. By Cretaceous time most of the uplands had been worn away, including the final obliteration by erosion of the land mass "Appalachia." Extensive shallow seas gradually flooded the continent.

B. Life of the Mesozoic

   1. Triassic.
      a. Ammonites abundant.
      b. Plants. Cycads and conifers.
      c. Vertebrates. Rise of the reptiles, primitive dinosaurs, and pterodactyls.
   2. Jurassic.
      a. Invertebrates.
         (1) Ammonites reached culmination; gastropods, pelecypods, and bryozoa abundant.
      b. Vertebrates.
         (1) Culmination of herbivorous dinosaurs, ichthyosaurs and mososaurs; first birds.
         (2) Development of primitive mammals.
      c. Plants.
   3. Cretaceous.
      a. Invertebrates. Decline and extinction of ammonoids,
         Culmination of bryozoa.
         Continued rise of pelecypods and gastropods.
         Spread of insects.
         Echinodermata important.
      b. Vertebrates.
         Culmination of carnivorous dinosaurs; extinction of all dinosaurs.
         Rise of archaic mammals and birds.
      c. Plants.
         Spread of flowering plants.
   4. Summary.
      The end of the Cretaceous, coming at the time of the Laramide Revolution, repeats the history of the Permian. The typically Mesozoic forms of life which had become overspecialized were wiped out and the dawn of the Cenozoic saw the beginnings of the types of life that are prevalent on the earth today.

EIGHTH WEEK

C. **Mesozoic Economic Deposits**

1. Oil geology.
    a. Nature and origin of oil.
    b. Conditions necessary for preservation of oil.
        (1) Source rock.
        (2) Reservoir rock.
        (3) Favorable structure.
    c. Oil prospecting methods.
        (1) Geophysical.
        (2) Structural.
        (3) Airplane mapping, photographs.
        (4) Wildcatting.
    d. Oil production methods.
    e. Oil fields of the United States.
        (1) Gulf Coast.
        (2) Pennsylvania and New York.
        (3) California.
        (4) Other producers.
    f. Foreign oil fields.
    g. Petroleum reserves.
2. Résumé of world mineral resources and their rôle in international politics.
    a. Minerals fundamental to industrial development.
        (1) Coal.
        (2) Iron ore.
        (3) Oil and gas.
        (4) Copper, zinc, lead, and aluminum.
        (5) Gold and silver.
        (6) Nitrates. Production almost entirely from Chile.
    b. Status of the nations in these resources.
        (1) United States. Well supplied with all fundamental minerals; deficient in manganese, nickel, and tin.
        (2) Great Britain. Abundant supplies of all fundamental minerals; potash and sulfur deficiencies.
        (3) France. Adequate supplies of iron, nickel, potash and aluminum only.
        (4) Italy. Adequate supplies only of aluminum, mercury, zinc, sulfur.
        (5) Japan. Adequate supplies only of copper and zinc.
        (6) U.S.S.R. Well supplied with coal, iron, gold, petroleum, manganese, platinum; new deposits of copper, lead, and zinc are being developed and may prove adequate.
    c. Recent political activities dictated in part by lack of essential minerals.
        (1) Japanese conquest of Manchuria for coal and iron.
        (2) Italian conquest of Ethiopia for what it may contain.
        (3) Germany's agitation for return of colonies.
        (4) The Treaty of Versailles and mineral resources.
        (5) Further examples.

EIGHTH WEEK

LABORATORY

Historical Geology

The laboratory on historical geology is directed toward an understanding of the nature of epeiric seas, and of the manner in which formations have been piled one upon another; later uplifted and eroded to produce the regional geology of today. Exercises to be worked out in the laboratory will be of the following types:

1. Construction of maps showing the change in the outline of North America by relative rises and falls of sea level.

2. Construction of a geologic map of the United States by building up the succession of rock deposition, period by period.

3. Interpretation of the history of a geologic period or of a geologic era by restored sections.

4. Construction of structure sections from data such as that obtained by field observations.

5. Given a series of facts about the geology of a region, to interpret the paleogeography, climate, and life conditions which existed during the time of deposition of the rocks.

FOR STUDY AND READING

ESSENTIAL
  Longwell, Knopf, and Flint, "Outlines of Historical Geology,"
    pp. 190-243;[1] "Outlines of Physical Geology," pp. 314-19.

SUGGESTED
  Agar, Flint, Longwell, "Geology from Original Sources," Ch. XIII.
  Egloff, "Earth Oil." Century of Progress Series.
  Grabau, "Historical Geology," Ch. XLI-XLIV.
  Lieth, "World Minerals and World Politics."
  Lucas, "Animals of the Past."
  Newman, and others, "The Nature of the World and Man," Ch. XI, first half.

---

[1] Page numbers refer to first edition; revised edition in preparation.

NINTH WEEK

AGE OF MAMMALS: THE CENOZOIC

To a geologist, the Cenozoic era which began forty million years ago represents modern earth history. The Cenozoic is commonly subdivided into the Tertiary which includes the first four epochs and the Pleistocene which is the closing epoch. This week's work will be concerned only with the events and life of Tertiary time.

During the Tertiary the mammals rose rapidly to prominence and evolved into many diverse forms. The face of the earth began to assume its present appearance. Only two major topographic changes occurred after the close of the Tertiary: the Cascadian-Himalyan mountain building and the Pleistocene continental glaciation.

A. <u>Cenozoic Era by Periods</u>

   1. Eocene.
      a. Overlap of seas along the Atlantic and Gulf Coastal Plain; marine deposits.
      b. Continental deposits containing many mammalian remains in enclosed basins of Rocky Mountain states.
      c. Marine invasion of northern California and thick volcanic deposits.
      d. Glaciation in western Colorado.
   2. Oligocene.
      a. Overlapping of seas along Gulf Coastal Plain.
      b. Marine invasion of northern California.
      c. Volcanic deposits in Washington and Oregon.
   3. Miocene.
      a. Overlap of seas along whole Atlantic and Gulf Coastal Plain.
      b. Marine deposits in southern California coast area.
      c. Extensive lava flows in Oregon and Washington (Columbia plateau lavas).
   4. Pliocene.
      a. Overlap of seas along outer margin of Atlantic and Gulf Coastal Plain.
      b. Marine deposits along Pacific coast margin.
      c. Lava flows in Oregon and Washington.
      d. Starting in Miocene, and continuing through Pliocene, the Cascadian Revolution along our west coast and the Alpine Revolution in Europe uplifted lofty mountain ranges. The Himalayas of Asia were also formed during this orogenic epoch.

B. <u>Areas of Cenozoic Rocks in the United States</u>

   1. Atlantic and Gulf Coastal Plains.
      Marine deposits along Atlantic and Gulf coasts lying on earlier Cretaceous deposits. Belted arrangements of outcrop with deposits of younger periods occurring nearer the present coast.

2. Early Cenozoic seas invaded central valley of California.
      Later ones were confined to a narrow belt along present ocean front.
   3. Mississippi embayment.
      Pleistocene and recent deposits of Mississippi River cover earlier Cenozoic deposits in this embayment.
   4. Gravel veneer over Great Plains.
   5. Glacial till deposited over Canada and much of north-central and northeast United States (Pleistocene).
   6. Sand and gravel basin-filling between block mountains of Great Basin.
   7. Thick lava flows beginning in Miocene time in Columbia Plateau.
   8. Igneous flows and intrusions in Colorado plateau and Pacific ranges.

C. Life of the Cenozoic

   1. Invertebrates.
      a. Gastropods.
      b. Pelecypods.
      c. Insects.
         These groups of animals, during the Cenozoic, and including recent time, became more numerous and occurred in greater diversity than at any time in the past.
      d. Echinoderms numerous.
      e. Bryozoa numerous.
   2. Vertebrates.
      a. Mammals.
         During the Cenozoic the mammals rose to great prominence and included all the types which we know today. They became adapted to the sea, the air, to climbing, burrowing, and other specialized activities.
      b. Reptiles.
         The larger reptiles (dinosaurs, etc.) became extinct.
         Numerous other specialized groups remained.
      c. Fish.
         Many of the modern types evolved and became numerous and diversified.
      d. Birds.
         All the modern forms of birds appeared, many groups becoming highly specialized.
   3. Plants.
      a. The flowering plants reached their highest degree of specialization during the Cenozoic. The grasses seem to have evolved during this time and greatly influenced animal evolution.
      b. Non-flowering plants (ferns, etc.) became much less important.

LABORATORY

The laboratory period for this week will be reserved for consultation, for laboratory demonstrations or for whatever new work or review of previous work seems most desirable at the time.

FOR STUDY AND READING

ESSENTIAL
  Longwell, Knopf, and Flint, "Outlines of Historical Geology,"
    pp. 244-91.[1]

SUGGESTED
  Grabau, "Historical Geology," Ch. XLV and XLVI.
  Newman and others, "The Nature of the World and Man," Ch. XI,
    last part.

---

[1] Page numbers refer to first edition; revised edition in preparation.

TENTH WEEK

THE GLACIAL AGE

The Pleistocene or last epoch of the Cenozoic brings the history of geology down to the present day. The continents had acquired their present structure and topography before they were in part overridden by ice sheets and mountain glaciers. Man had certainly made his appearance and it is likely that the encroachment of the ice upon the domains which he was inhabiting forced him to invent ways of living which would enable him to meet these harsher conditions. It is not unlikely that some of the sterner qualities of the human race have come from this struggle.

A. **History of Pleistocene Time.** Continental glaciation

1. In North America.
    a. Several centers of dispersion. Keewatin, Labrador, Newfoundland, and Cordilleran.
    b. Area covered. Equivalent to present Antarctic ice sheet. Approximately to Ohio and Missouri Rivers.
    c. Stages of glaciation (alternating with interglacial stages). Kansan (Jerseyan), Nebraskan, Illinoian, Iowan, and Wisconsin. Oscillations in each stage.
2. In Europe, Asia, and South America.
    a. Center of dispersion in Scandinavia.
    b. Area covered. Position of terminal moraine, in England, Holland, Germany, Poland, and Russia.
    c. Stages. Evidences of oscillations.
3. Effects of glaciation.
    a. Erosion. Removal of soil; striae, grooves, <u>rôches moutonnées</u> rock basins, deepening of valleys as in Hudson River gorge.
    b. Deposition.
        (1) Moraines of several types. Terminal, recessional, interlobate, ground.
        (2) Moraine pattern determined by basins and glacial lobes in Great Lakes area and in north Germany.
        (3) Specific examples of the Terminal Moraine. Martha's Vineyard, Nantucket, Long Island, Staten Island, the Great Lakes region, South Dakota, Montana. East Prussia and Denmark. Morainal belts.
        (4) The driftless area. Wisconsin and Iowa.
        (5) Ground moraine. Distinction between transported and residual soil. Varying thickness and character.
        (6) Drumlins. Methods of origin. Drumlin swarms: Boston Bay, Wisconsin, New York State, and Sweden.
        (7) Fluvio-glacial deposits. Gravel, sand, clay.
            (a) Outwash. Long Island, Middle West, glacial rivers; Denmark, Prussia.
            (b) Eskers. Maine, Scandinavia.
            (c) Kames. Numerous examples. Kettle holes.

(8) Lake Deposits. Varved clays. Use in determining time
   since retreat of ice.
c. Drainage changes due to glaciation.
   (1) Blocking of valleys to form lakes. The Canadian Lake
      area, Wisconsin, Minnesota, New England, Finland.
   (2) Pro-glacial or marginal lakes. Lake Agassiz, Lake Passaic
      (now extinct).
   (3) Lakes formed in tributary valleys by alluviation of main
      stream. Allegheny River, Ohio River, Idaho.
   (4) Temporary disturbance of streams. Columbia River and
      Grand Coulee, Missouri River and Shonkin Sag.
   (5) Origin of waterfalls. Niagara.
   (6) Glacial outlets.
d. Post-glacial effects.
   (1) Warping. Great Lakes region, Great Basin, Scandinavia.
   (2) Relation to coral reef theory.
e. Economic effects of continental glaciation.
   Soils, swamps, lakes, water power.
4. Cause of glacial period. Several theories.
   a. Terrestrial.
      (1) Emergence of land; elevation of land; change of ocean
         currents.
   b. Atmospheric.
      (1) Change in volume of carbon dioxide. Excess of volcanic
         ash in the atmosphere.
   c. Astronomic.
      (1) Variations in solar energy; precession of equinoxes.
5. Comparison with earlier glacial periods in Pre-Cambrian and
   Permian time.
6. Effect of glacial period upon plant and animal life.

B. History of Pleistocene Time. Alpine glaciation (synonymous with
   valley, mountain, and local glaciation).

1. In North America.
   a. Former extent of glaciers in Rocky Mountains, Sierra Nevada,
      and other western ranges.
   b. Alpine glaciation in eastern United States. White Mountains,
      Mount Katahdin, Adirondacks.
2. In Europe.
   a. Alpine glaciation in Scandinavia, Scotland, Wales, the Alps,
      Pyrenees, Carpathians.
3. On other continents.
   a. The Andes, New Zealand, the Himalayas.
4. Effects of Alpine glaciation.
   a. Erosional effects.
      (1) Cirques, rock basins, glacial troughs with hanging valleys,
         arrêtes, matterhorn peaks. Biscuit-board topography.
         Stages in glacial erosion.
      (2) Asymmetry of glaciation in Rockies and White Mountains.
      (3) Description of typical examples. Yosemite, Glacier Park,
         Rocky Mountains, Uintas.
   b. Depositional effects.
      Moraines. Terminal, lateral, medial, horseshoe shaped.
      Rocky Mountain National Park.
   c. Lakes due to Alpine Glaciation. Tarns, Finger Lakes.

       d. Fiords. Due to erosion below sea level. Not necessarily to submergence.
       e. Economic effects of Alpine glaciation. Railroads, agriculture, cities.
  5. Piedmont Glaciers.
       Alaska, Italy, Bavaria, Glacier Park.
  6. Contrast between stream erosion and glacial erosion.
       Stream-formed valleys. V-shaped, full-bodied forms, deep soil cover, few rock ledges, accordant streams (Playfair's Law).
       Glaciated Valleys. U-shaped, angular forms, rock ledges, hanging valleys, and waterfalls.

LABORATORY

The laboratory exercises covering Pleistocene and present-day aspects of the earth involve a study of land forms produced by the geological processes of glaciation, stream action, waves, wind, and weathering, as well as forms resulting from organic activity. They involve also the larger structural types of plains and plateaus and the various types of mountains.

It is here that the student working indoors finds himself at a disadvantage. It is only fair to say that landscapes can be understood best only by direct observation. The chemist and the physicist are fortunate in being able to bring into the laboratory the things they are studying, but the geological student must resort to maps, models, and diagrams in order to gain a comprehension of land forms.

Acquaintance with topographic maps will be made first through the topographic sheets of the U. S. Geological Survey and representative examples of foreign maps. This will be followed by selected examples of maps to illustrate the various land forms under discussion.

1. <u>Topographic Maps of United States and Foreign Countries</u>.

    a. Name or number; publisher; index.
    b. Scale.
      (1) Fractional
      (2) Graphic.
      (3) Verbal.
    c. Relief.
      (1) Contours.
      (2) Hachures.
      (3) Layering.
    d. Drawing a Profile.
      (1) To natural scale.
      (2) With vertical exaggeration.
    e. Symbols and colors used. Conventional signs.
    f. Map reading vs. map interpretation.

       <u>Problems</u>. Assignments in constructing maps, drawing profiles, and related topics on maps of various kinds.

2. <u>Representative Maps Illustrating Glacial Topography</u>.

    a. Chief Mountain, Montana.

TENTH WEEK   39

   b. Mt. Rainier National Park.
   c. Aletschgletscher, Switzerland.
   d. Narvik, Norway.
   e. St. Croix Dalles.
   f. Quebec, Canada.

3. Study exhibits of wall maps, pictures, and diagrams illustrating
   Pleistocene glaciation in various parts of the world.

4. Study of models.

   The following 18 colored models illustrating all types of
structure and topography will be studied during this week and the next
two weeks. As each model illustrates several different topics it is
not practicable to classify them into three groups.

   Accompanying each model is the topographic map of the same
area on the same scale, together with questions and suggestions for
study.

   1. Delaware Water Gap, Pennsylvania. Dipping beds.
   2. Boothbay, Maine. Submerged coast.
   3. Yosemite Valley, California. Glaciation.
   4. Niagara Falls, New York. Slightly dipping beds and retreat
      of falls.
   5. Oceanside, California. Marine terraces.
   6. Everett, Pennsylvania. Folded Mountains.
   7. St. Louis, Missouri. Plateau.
   8. Mt. Shasta, California. Glaciated volcano.
   9. Mt. Taylor, New Mexico. Volcanic necks.
   10. Cucamonga, California. Alluvial fans.
   11. Bellefonte, Pennsylvania. Folded Mountains.
   12. Denver, Colorado. Hogbacks.
   13. Chief Mountain, Montana. Glaciation.
   14. Spanish Peaks, Colorado. Volcanoes and dikes.
   15. Kaaterskill, New York. Plateau. Stream capture.
   16. Henry Mountains, Utah. Laccolithic domes.
   17. Flagstaff, Arizona. Mature volcanoes.
   18. Mount Lyell, California. Young volcanoes.

5. Model of Chart B of "Panorama of physiographic types." This large
   colored model embraces numerous types of land forms and geological structures, all of which are described in an accompanying
   pamphlet.

FOR STUDY AND READING

ESSENTIAL
   Longwell, Knopf and Flint, "Outlines of Historical Geology," pp. 292-
   306;[1] "Outlines of Physical Geology," pp. 85-103, 103-12; "Panorama
   of Physiographic Types," par. 59 and 60.

---

[1]Page numbers refer to first edition; revised edition in preparation.

SUGGESTED
  Geike, "The Great Ice Age."
  Grabau, "Geology," Part II, pp. 885-91 on the causes of glaciation,
    also pp. 864-85 on the ice age in North America and Europe.
  "Panorama of Physiographic Types," exercises following par. 60.
  Thwaites, "Outline of Glacial Geology."

ELEVENTH WEEK

SCULPTURING AGENTS OF THE MODERN LANDSCAPE

The present appearance of the modern landscape is the result of the destructive geological forces, such as weathering, streams, glaciers, waves, winds, and organisms, acting upon certain larger constructional forms, such as plains, plateaus, and the various types of mountains (dome, block, folded, complex, and volcanic).

This week's work has to do with the work of the destructive forces. Next week takes account of the constructional forms, which are the elements of continental structure. The following week treats of the continents themselves and their present-day plan.

The time since the last advance of the ice is usually termed "Recent," and the length of this interval is estimated to be from 20,000 to 80,000 years. Some of the Pleistocene interglacial intervals were as long as Recent time, and it is altogether possible that the present is but a congenial interglacial time. Even so, continental glaciation has not entirely receded from the earth, for somewhat more than five million square miles of the earth's surface still lie beneath perpetual snow and ice. Many of the details of present-day landscapes were formed before the glacial age. Glaciation is therefore to be thought of as merely one of the sculpturing forces which has produced present-day topography.

A. **Weathering and Soils**

1. Chemical effects of weathering.
    a. Carbonation and hydration, exemplified by change of orthoclase to kaolin and potassium carbonate thus:

    Orthoclase + Water + Carb. Diox. = Kaolin + Qtz. + Pot. Carb.
    $2KAlSi_3O_8 + 2H_2O + CO_2 = H_4Al_2Si_2O_9 + 4SiO_2 + K_2CO_3$

    Feldspar (orthoclase) affected by surface water carrying CO_2, and by hot waters from below.
    Clay (kaolin) is a stable mineral on the earth's surface. Occurs everywhere; glacial deposits, river alluvium, residue from limestone, lake beds, marine shales. White when pure. Used for fine pottery. Usually red, yellow, or orange because of iron. Black with manganese.
    Quartz (silica) carried in solution; deposited around sand grains in sandstone to form quartzite; also in quartz veins, or as siliceous sinter and geyserite. Not very soluble but total amount removed is great.
    Potassium carbonate. Like most carbonates very soluble. Extremely important in making potassium available for plant use.
    b. Carbonation of copper to form malachite and azurite, e.g., Columbia roofs.

c. Hydration represented by formation of limonite, or bog ore, $2Fe_2O_3 \cdot 3H_2O$. Also Bauxite, $Al_2O_3 \cdot 2H_2O$.
d. Oxidation. Represented by the several iron oxides, hematite, Fe_2O_3, magnetite, Fe_3O_4, and the hydrated oxides.
2. Mechanical effects of weathering.
    a. Effect of heat.
        Exfoliation, most effective on massive rocks, basalt.
        Exfoliation domes. Yosemite; Brazil.
        Disintegration of granite, due to different coefficients of expansion of minerals. Natural gravels, disintegration boulders. Rocky Mountains.
    b. Effect of cold. Ice in joints.
        Formation of talus at foot of cliffs. Palisades.
3. Solifluxion. Mud flows. Creep of soil. Landslides.
4. Organic influences.

B. **Streams**

Distinction between the *ages* of a stream and the *stages* in the development of a region.

1. Young streams (excess of energy). Colorado and Yellowstone rivers.
    a. Downward cutting predominant, but accordant junctions. (Playfair's Law).
    b. V-shaped valleys. Narrow valley floor.
    c. Waterfalls, cascades, rock ledges.
        Retreat of falls. Horizontal strata.
        Wearing down of falls.
    d. Lakes. Clear water.
    e. Stream capture.
        (1) By streams flowing on lower level.
        (2) By streams flowing along weak belt.
        (3) By streams cutting laterally.
2. Mature streams (graded; no excess of energy), with examples of the following characteristics:
    a. Lateral cutting.
    b. Flood plain. Natural levees.
    c. Meanders and cut-offs.
    d. Deltas and alluvial fans.
3. Rejuvenated streams.
    a. Incised meanders.
    b. Terraces.
4. Classification of streams.
    a. Genetic: Consequent, subsequent, obsequent, resequent, insequent. Superposed, antecedent.
    b. Pattern: Dendritic, radial, trellis, rectangular, annular.

C. **Waves**

1. Shorelines of submergence.
    a. Youth. Irregular coast. Stacks, arches, caves. Truncated headlands, various types of spits, (simple, recurved, compound, complex); bars (bayhead, baymouth, midbay, bayside, etc.); and beaches. Islands and tombolos (simple and complex).
    b. Maturity. Headlands cut back to heads of bays. Graded shore profile.

       c. Old age. Wave-cut peneplane.
   2. Shorelines of emergence.
       a. Youth. Simple or straight coast. Shallow offshore. Wave nip,
          barrier bars, beach ridges, tidal deltas. Lagoon.
       b. Maturity. Bar thrown back on mainland. False evidence of
          recent emergence.
       c. Old age.
   3. Neutral shorelines.
      Delta coasts.
   4. Fault shorelines.
   5. Compound shorelines.

D. **Wind**

   1. Erosional activity. Deflation. Abrasion.
   2. Transportation. Dust storms. Amount transported. Distance.
   3. Deposition. Loess. Dunes.
   4. Relation of wind deposits to glaciation in Europe, New Zealand,
      and United States.
     Relation of wind deposits to arid areas in United States, China,
      and South America.
     Relation of wind deposits to river flood plains in United
      States.
     Relation of wind deposits to sea coast.

E. **Underground Processes**

   1. Artesian wells.
     Ground water table. Perched water tables. Flowing wells.
     Structural conditions. Dipping beds, alluvial fans, glacial
      till, joints.
   2. Caves and underground drainage.
     Sink holes. Valley sinks. Natural bridges. Caves. Dripstone: stalactites, stalagmites. Karst topography. Typical
      cave regions.
   3. Springs. Mineral springs. Hot springs.
     Relation to structure, joints, and faults.
   4. Geysers.
     Relation to faults and grabens.

F. **Organisms**

   1. Coral reefs.
      Barrier reefs; atoll reefs.
      Theories of reef formation:
     a. Subsidence theory of Darwin.
     b. Glacial control theory of Daly.
     c. Theory of stationary levels of Vaughan and Murray.

LABORATORY

1. **Representative Maps Illustrating Streams.**

   a. Canyon quadrangle. Young stream, canyon.
   b. Elk Point quadrangle. Mature stream, flood plain, terraces.

c. Lancaster quadrangle. Young streams. Mature region.
   d. Memphis quadrangle. Mature river, meander scars, natural levees.
   e. Mannheim, Germany. Mature stream, flood plain, old meander scars.
   f. Windermere, England. Mature stream, delta, glacial lakes.

2. <u>Representative Maps Illustrating Wave Action.</u>
   a. Boothbay, Maine. Shoreline of submergence.
   b. Barnegat, New Jersey. Shoreline of emergence, primarily.
   c. Ocean City, Maryland. Shoreline of emergence.
   d. Coast Chart No. 1215. Approaches to New York. Shoreline of emergence and also of submergence.

3. <u>Representative Maps Illustrating Wind Action.</u>

   a. Ogalalla, Nebraska. Dunes and loess.

4. <u>Representative Maps Illustrating Underground Processes.</u>

   a. Interlaken, Florida. Sink holes.

5. <u>Representative Maps Illustrating Work of Organisms.</u>

FOR STUDY AND READING

ESSENTIAL
   Longwell, Knopf, and Flint, "Outlines of Physical Geology," Ch. III, omitting pp. 59-65, 66-76, 113-43, 214-19; "Panorama of Physiographic Types," pars. 52-57 on streams, 58 on underground water, 62-67 on waves.

SUGGESTED
   Davis, "Physical Geography," pp. 222-83.
   Hobbs, "Earth Features and Their Meaning."
   Johnson, "Shore Processes and Shoreline Development."
   Tarr and von Engeln, "New Physical Geography."

TWELFTH WEEK

THE LARGER ELEMENTS OF THE MODERN LANDSCAPE

These elements are sometimes termed the constructional land forms. Their essential character is determined by their structure. They are the blocks or units which make up the continents. If the structure is simple and undisturbed the region is a plain or plateau. If the structure is disturbed the region is a mountain area, the particular kind of mountain being determined by the particular kind of structure. In their development these units pass through a series of stages depending upon the amount of erosion they have suffered. To these stages the terms "young," "mature," and "old" are applied. A region is young as long as much of the initial surface remains. It is mature when it is all cut up into hills and valleys, and old when reduced to a peneplane.

A. <u>Plains</u>

    1. Coastal plains.
       a. Narrow coastal plains. Scotland, Maine.
       b. Belted coastal plains.
          Inner lowland, cuestas, (inface and backslope).
          Outer lowland, fall line.
          Inliers, outliers. England, France, Alabama, Texas, Wisconsin.
       c. Submerged belted coastal plains. Long Island, Great Lakes region, Baltic Sea region.
    2. Interior plains.
       a. Young. Lake Agassiz basin.
       b. Mature. Iowa, Missouri.
       c. Old. Plains of Russia, Montana.
    3. Tilted structures. Dip; strike. Triassic lowland.

B. <u>Plateaus</u> (Plateaus are regions of high relief; plains are regions of low relief)

    1. Young plateaus. Deep canyons, Columbia, Colorado plateaus.
    2. Mature plateaus.
       Coarse textured (Catskills); medium (western Pennsylvania); fine (West Virginia).
    3. Old plateaus. Mesas and buttes. Montana, North Dakota.
    4. Rejuvenated plateaus, Grand Canyon region.
    5. Warped plateaus. Domes and basins.
       Interior plateaus. Kentucky and Tennessee.
    6. Broken plateaus.
       Grand Canyon region. Central Washington.

C. <u>Dome Mountains</u>

    1. Cause of domes. Batholiths, laccoliths, warping.

2. Young domes.
    Radial drainage.
    Salt domes of Louisiana and Texas.
3. Mature domes.
    Annular drainage.
    Crystalline core exposed. Black Hills.
    Crystalline core not exposed. Weald, England.
    Hogbacks, water gaps, wind gaps. Stream capture.
4. Subsidiary and parasitic domes.
5. Economic aspects.

D. Block Mountains and Faulted Structures

1. Simple and multiple faulting. Fault splinters.
2. Young block mountains.
    a. Fault scarp and back slope. Oregon.
3. Mature block mountains:
    a. Straight base line truncating structure.
    b. Triangular facets.
    c. Evidence of recent faulting.
    d. Alluvial fans.
    e. Springs.
    f. Examples: Nevada and Utah.
4. Old block mountains.
    a. Rock pediments--conoplains.
    b. Examples: Nevada, Arizona, New Mexico.
5. Grabens and horsts. Rhine Graben. Death Valley. Harz Mtns.
6. Fault scarps and fault line scarps. Grand Canyon Region.
    Ramapo Fault.
7. Fault control of stream pattern. Adirondacks.
    Fault control of lakes. Sweden.
8. Thrust faults. Lewis Thrust.

E. Folded Mountains

1. Initial folding and initial consequent stream system.
2. Maturity.
    a. Reversal of topography.
    b. Trellis stream pattern. Consequent, subsequent, obsequent,
       and resequent streams.
    c. Anticlinal, synclinal, and monoclinal mountains.
    d. Anticlinal, synclinal, and monoclinal valleys.
3. Pitching anticlines and synclines.
    a. Contrast in outcrops and in topography between the two types.
    b. Zigzag ridges.
4. Overturned folds.
5. Anticlinoria; synclinoria.

F. Complex Mountains

1. Types of complexity.
2. Young, mature, old, rejuvenated.

G. Volcanic Forms

1. Eruptive types.

TWELFTH WEEK

    a. Young. Cones, craters, compound cones. Breached craters.
       Lava flows. Blocked valleys. Types of ejecta. Bombs.
    b. Maturely dissected volcanoes.
       Stream and glacial erosion.
    c. Old volcanoes.
       Volcanic necks. Dikes.
    d. Wrecked volcanoes.
       Calderas.
       Katmai, Krakatoa, Crater Lake.
2. Lava domes.
    a. Young domes.
       Collapsed basins. Fissures. Incipient drainage. Faults.
    b. Maturely dissected domes.
    c. Old domes.
    All of these stages represented by the Hawaiian volcanoes. Also in Iceland, Reunion, etc.
3. Lava flows and lava plateaus.

LABORATORY

1. <u>Representative Maps Illustrating Plains and Plateaus.</u>

    a. Conde, South Dakota. Very young plain.
    b. Watrous, New Mexico. Young plateau.
    c. Oceana, West Virginia. Mature plateau.
    d. Latrobe, Pennsylvania (structural section sheet). Warped plateau.
    e. Bright Angel, Arizona. Maturely dissected plateau.

2. <u>Representative Maps Illustrating Dome Structures.</u>

    a. Watrous, New Mexico. Small eroded dome.
    b. Oregon Basin, Wyoming; Meteetse, Wyoming. Large eroded domes.

3. <u>Representative Maps Illustrating Block Mountains and Faults.</u>

    a. San Pedro, New Mexico. Young block mountains.
    b. Ramapo, New Jersey. Fault-line scarp.

4. <u>Representative Maps Illustrating Folded Mountains.</u>
    a. Harrisburg, Pennsylvania. Pitching syncline.
    b. Everett, Pennsylvania. Pitching anticline.

5. <u>Representative Maps Illustrating Complex Mountains.</u>

    a. Philipsburg, Montana (structural section sheet).
    b. Three Forks, Montana (structural section sheet).
       Complex parts of the Rocky Mountains involving all kinds of rocks.

6. <u>Representative Map Illustrating Volcanoes.</u>

    a. Naples, Italy. Vesuvius, a young volcano.

FOR STUDY AND READING

ESSENTIAL
Longwell, Knopf, and Flint, "Outlines of Physical Geology," pp. 237-43, 281-85; "Panorama of Physiographic Types," Sections 14-51. Identify all types and features upon the charts. "Topographic and Structural Geology," Sections 24-30.

SUGGESTED
Davis, "Physical Geography," pp. 113-221.
See also list of references at end of Syllabus.

THIRTEENTH WEEK

THE PRESENT WORLD STAGE

To know something about the various elements which make up a landscape and to appreciate the numerous details produced by the several sculpturing agents is not enough, if one essays to have an understanding of the world stage and the movements of man upon it.

The world stage is like a mosaic with its various parts fitted and integrated together into a pattern which has a meaning. The different units are related to each other. Each continent is made up of a number of elements or provinces, each having its own peculiar structure, topography, and resulting natural resources, but which at the same time bears some relationship to those which are adjacent to it.

The merest outline and introduction to this vast subject will be attempted this week.

The purpose of this presentation is to give the student a concrete picture of the land areas of the world, as they actually are; to show him that the present distribution of landscapes is the result of all the past history of the earth, and that the continents are highly organized entities having very definite patterns. A background of this kind is invaluable to the student of human culture. It is obvious that man, in his wanderings over the earth's surface, both in war and in peace has been profoundly influenced by the character and distribution of topographic forms. The small insight which this course can provide is quite inadequate even to suggest the possibilities which this field of study has to offer the student of history, archaeology, economics, geography, and other subjects in the field of the humanities. This background is absolutely fundamental to the natural sciences which are concerned with the distribution of living forms and of mineral resources.

The lectures this week will present a simple analysis of the physical features of the different continents. Each continent will be divided into several large units and these in turn will be subdivided into the main physiographic provinces. An ideal physiographic province is one which has the same geologic structure throughout and has reached the same stage of erosion everywhere. In such cases geologic provinces and physiographic provinces coincide. But most physiographic provinces vary considerably from place to place and sometimes the boundaries between them must be drawn in an arbitrary manner. If each physiographic province into which a continent is divided is made as simple and clean-cut a unit as possible then it will be necessary to distinguish large numbers of provinces. But if it is desired to break up a continent into a few large sections, then each one of these sections will necessarily have a large number of details and many variations from the simple type.

A. North America

   1. United States.
      a. Laurentian upland.
         Superior upland. Adirondack Mountains.
      b. Atlantic plain.
         Continental shelf. Coastal plain.
      c. Appalachian highlands.
         Older Appalachians and New England.
         Folded Appalachians.
         Appalachian plateaus.
      d. Interior lowlands.
         Interior low plateaus.
         Central lowland.
         Great Plains.
         Ozarks and Ouachitas.
      e. Rocky Mountains.
         Northern and Southern Rockies.
      f. Intermontane plateaus.
         Columbia plateau.
         Colorado plateau.
         Basin and range province.
      g. Pacific mountain system.
         Sierra Nevada and Cascade Mountains.
         Coast ranges and associated troughs.
   2. Canada and Alaska.
   3. Mexico.
   4. Middle America.
      Central America and Caribbean.

B. South America

   1. The Brazilian highland and associated features.
   2. The Guiana highland.
   3. The Andean system.
   4. The Orinoco plain.
   5. The Amazon plain.
   6. The Parana - Paraguay plain.
   7. The Patagonian plateau.

C. Europe

   1. The Northwest highlands, and associated lowlands.
   2. The Great Central Plain.
   3. The Central massives and associated lowlands.
   4. The Alpine system and associated massives and lowlands.

D. Asia

   To be studied as class problems.

E. Africa

   To be studied as class problems.

F. **Australia**

To be studied as class problems.

LABORATORY

The laboratory period will be devoted to a review of the physiographic provinces of the different continents. The following material will be used:

1. Geological map of North America.
2. Geological map of United States.
3. Geological maps of representative states.
4. Geological map of Europe.
5. Geological maps of European countries.
6. Geological map of the world.
7. Wall maps of the continents.
8. Selected topographic maps of the different countries of the world.
9. Models and pictures.

It can hardly be expected that the student will acquire a very full knowledge of all of these regions but it is hoped that a fairly concrete picture can be established as the basis for further learning when the need and opportunity arises. And it is expected that the student will become acquainted with some of the sources of information to which he can turn for help.

FOR STUDY AND READING

ESSENTIAL
  Physiographic Diagram of the United States.
  Physiographic Diagram of Europe.
  Physiographic Diagram of South America.
  Specific assignments will be made to each student.

SUGGESTED
  Blanchard and Crist, "A Geography of Europe."
  Bowman, "Forest Physiography" (of the United States).
  Brooks, "Europe Including the British Isles."
  Fenneman, "Physiography of the Western United States."
  "Geographie Universelle."
  Loomis, "Physiography of the United States."
  Lyde, "Continent of Europe;" "Continent of Asia."
  Partsch, "Central Europe."
  Valkenburg and Huntington, "Europe."

FOURTEENTH WEEK

ADVENT OF MAN

A. Man, Origin and Early Types

   1. Classification of man.
      a. Grouped under Primates (lemurs, monkeys, and apes) by Linnaeus, middle eighteenth century.
      b. Close relation to the anthropoid apes (gorilla, chimpanzee, and orang), T. H. Huxley, 1862.
      c. Evidence of relationship from anatomy, embryology, pathology, and comparative psychology.
      d. Descent from tree-living ancestors.
   2. Antiquity of the Primates.
      a. Rarity of fossils.
      b. Occurrence in Tertiary rocks of North America, Europe, southern Asia, and East Africa.
   3. Divergence of man, adapted for bipedal life on the ground, from anthropoid stock, adapted chiefly for arboreal life.
      a. Probably in Miocene or early Pliocene.
      b. Origin in Asia or Africa, probably southern Asia; certainly not in America or Australia.
      c. No known fossil ape can be definitely stated to be a direct ancestor of man.
   4. Human paleontology.
      a. Earliest skeletal remains of early Pleistocene age.
      b. Successive types progressively higher.
         1. Pithecanthropus erectus, early or mid-Pleistocene, Java.
         2. Sinanthropus pekinensis, early Pleistocene, China.
         3. Eoanthropus dawsoni, Pleistocene, England.
         4. Homo (Palaeanthropus) heidelbergensis, mid-Pleistocene, Germany.
         5. Homo neanderthalensis, late Pleistocene, Europe and Palestine.
         6. Homo rhodesiensis, Africa (probably Pleistocene).
         7. Homo javanensis (Solo man), Pleistocene, Java.
         8. Homo sapiens, late Pleistocene, Europe, Java, Africa, Australia. Earliest men in America late migrants from eastern Asia.

B. Beginnings of Human Culture

   1. The background of culture.
      a. Transmission of learned behavior among animals.
      b. Social heredity as an aid in rapid adaptation to environment.
      c. Human culture as the culmination of a general mammalian tendency.
   2. Factors distinguishing culture from the social heredity of animals.
      a. Use of tools.

              b. Use of fire.
              c. Language.
         3. Evidence for existence of culture in species other than our own.
              a. The Taungs species.
              b. The Peiping species.
              c. Probability that by the time evolving man became Homo sapiens
                 he already had a fairly extensive culture equipment.
         4. Evidence for the reconstruction of human culture beginnings.
              a. Direct evidence (archaeology).
                 (1) Why such evidence is scanty.
                 (2) Nature of the evidence and what can be deduced from it.
              b. Inference from living groups of low culture.
                 (1) Dangers: All living cultures have evolved far beyond
                     original condition.
                 (2) Limitations imposed by a simple technology.
                 (3) Culture patterns common to such groups.
         5. The beginnings of technology.
              a. The first tools.
                 (1) Types.
                 (2) How stone implements are made.
              b. The use of fire.
                 (1) Fire in nature.
                 (2) Primitive fire making methods.
                 (3) Probability that use of fire long preceded ability to
                     make it.
         6. Influence of these advances on human life.
              a. Opening of new environments to human occupation.
                 (1) Tools and the food supply.
                 (2) Fire and the food supply. How fire increased supply of
                     vegetable foods.
              b. Resulting wide distribution of Homo sapiens and increase in
                 numbers.
         7. The beginnings of language (purely conjectural).
         8. The beginnings of society.
              a. Archaeological evidence.
              b. The general primate social patterns.
              c. The universal human social units.
                 (1) The family (physiological basis, factors giving continuity
                     to human matings, etc.)
                 (2) The local group.
              d. The first steps in specialization.
              e. The beginnings of social control.
         9. The beginnings of religion.
              a. Archaeological evidence.
                 (1) The earliest burials.
                 (2) The cave-bear cult of Switzerland.
              b. Evidences from modern primitives.
                 (1) Dreams and animism.
                 (2) The medicine man.
        10. Composite picture of the life of our first ancestors.

    C. Distribution of Man.   Race and Culture.

         1. Essential unity of man.
              a. Proofs that all living human groups belong to a single species.
                 (1) Production of fertile hybrids on crossing.

(2) Range of human variation compared with that of other mammalian species, especially other primate species.
2. Causes of present differentiation.
   a. Early human migration.
      (1) Wide range of environments made available to species by knowledge of tools, fire, etc.
      (2) Presumably rapid spread of species over Old World.
   b. Influence of natural environment on physical type.
      (1) Light intensity and pigmentation.
      (2) Other possible influences.
   c. Social selection.
      (1) Varying ideas of beauty.
      (2) Influence of selective mating.
      (3) Occasional direct elimination of disapproved types.
   d. Primitive social conditions as an aid to the fixation of physical types.
      (1) Limited size of primitive groups for economic reasons.
      (2) Close inbreeding of primitive groups.
3. Living races.
   a. Definition of race: A group of individuals united in blood and heredity.
   b. Methods of racial determination.
      (1) Lack of records showing actual ancestry.
      (2) Significance of physical resemblances.
      (3) Method of distinguishing between groups of mixed and of unmixed heredity.
   c. Extreme rarity of unmixed human groups.
4. Racial classifications.
   a. Tentative nature of all racial classifications.
      (1) Classifications based on a limited number of criteria selected for purpose.
      (2) Improbability that all members of any one of great racial divisions now recognized actually have a common ancestry distinct from that of the rest of mankind.
      (3) The Caucasoid, Mongoloid, and Negroid divisions of mankind.
5. The significance of race.
   a. Social significance as distinct from intrinsic significance.
      (1) Race as a determinant of social status.
      (2) Causes leading to use of race as a determinant of status.
      (3) Rationalization of social inequality of races.
   b. Intrinsic significance of race.
      (1) Question of physical superiority or inferiority.
         Such superiority a relative matter. Particular racial types superior in particular environments. Cf. Negro and white in Europe and West Africa.
      (2) Question of mental differences crux of problem.
         Have all living groups sufficient intelligence to become civilized?
6. Racial differences in mentality.
   a. Difficulties of exact methodological approach.
      (1) Tests devised on basis of European culture put individuals of other cultures at great disadvantage.
      (2) Culture factors least important in sense tests. Such tests show no marked racial differences.

FOURTEENTH WEEK

      b. Conclusions from informal observation.
         (1) Uniform unwillingness of experienced field workers to admit that native groups they know well are inferior to whites in intelligence or differ in mental processes.
      c. The evidence of history.
         (1) Easy assumption of new culture elements by all racial groups.
         (2) Way in which different races have taken the lead in cultural advance in different periods.
      d. Conclusion that all the large racial divisions of mankind are approximately equal in their ability both to carry and enrich civilization.
  7. Race and culture today
      a. Improbability that racial differences will be eliminated.
      b. Elimination of technological differences in cultures going on rapidly. Cf. Japan.
      c. Ultimate loss of white technological superiority and with it end of white world domination.
      d. Social discriminations on a racial basis as a cause of friction.
      e. Desirability of a change in attitude of whites toward other racial groups.

LABORATORY

     The purpose of this laboratory period is to demonstrate some of the relations between man and his environment. Use will be made of maps, pictures, models, and charts.

1. **Man and His Distribution.**

      a. The distribution of races, nations, and language.
      b. Distinction between race and language.
      c. Regional aspects of human culture.

2. **Man and His Natural Environment.**

      a. Factors influencing location of
         (1) Cities.
         (2) Transportation routes.
      b. Factors influencing character of activities.
         (1) Agriculture.
         (2) Mining.
         (3) Fishing.
      c. Factors influencing the thought of man.
         (1) Philosophy.
         (2) Religion.

3. **Man and Natural Resources**

      a. Exhaustible resources.
         (1) Coal and minerals
         (2) Lumber.

            (3) Soil.
            (4) Animal life.
        b. Replenishable resources.
            (1) Lumber.
            (2) Animal life.
        c. Inexhaustible resources.
            (1) Water power.
            (2) Wind power.
            (3) Tidal power.
            (4) Sun power.
            (5) Nitrates in the air and certain minerals of great abundance,
                e.g., iron, salt, clay, silica.
        d. Limiting factors in population density.
        e. The place of science, economics, and politics in regulating
            man's affairs.

FOR STUDY AND READING

ESSENTIAL
   Schuchert, "Outlines of Historical Geology," Ch. XXI, pp. 307-23.

SUGGESTED
   Hooton, "Up from the Ape."
   Linton, "Study of Man," Ch. II-III.
   MacCurdy, "Human Origins," Vol. I.
   Quennell, "Everyday Life in the Old Stone Age."
   Ripley, "Races of Europe."
   Sollas, "Ancient Hunters and Their Modern Representatives."

FIFTEENTH WEEK

THE DISTINCTIVE ASPECTS OF HUMAN CULTURE

A. **The Distinctive Aspects of Human Culture**

   1. The factors which make culture as we know it possible.
      a. Language.
      b. Organized social life.
      c. Distinctive qualities of the human mind.
   2. Language and culture.
      a. The rôle of language in culture transmission.
         (1) Language makes possible complete transmission of culture.
             Behavior patterns adapted to unusual situations.
             Abstractions as knowledge, ideas, etc.
   3. Social life and culture.
      a. Organized social life makes possible specialization.
      b. The group, not the individual the culture-bearing unit.
      c. Effects of specialization on culture content.
   4. Human mind and culture.
      a. Superior learning ability of man.
      b. Restless energy of human mind.
         Restlessness general Primate characteristic.
         General human unwillingness to let well enough alone.
   5. The distinctive aspects of culture.
      a. Extreme richness of content.
      b. Tendency toward progressive enrichment.
      c. Increasing divorce from factors of necessity.
   6. Richness of human culture content.
      a. Human culture incomparably richer than social heredity of any other species.
      b. Dominant principle that of producing adaptive changes in environment rather than adaptation to environment.
      c. Infinite possibilities of enrichment opened by this approach.
   7. Tendency toward progressive enrichment.
      a. Historic evidence for such progressive enrichment.
      b. The causes of culture growth.
         (1) The stimulus of new needs (of secondary importance).
         (2) Mental restlessness.
      c. The processes of culture growth.
         The inventor.
         Social acceptance or rejection and its causes.
         Cultural borrowing (diffusion).
         Irregular rates of culture advance.
         Disharmonic developments.
   8. Culture and necessity.
      a. Development of social heredity as an adaptive mechanism.
      b. Minimal culture content which would suffice to insure survival of our species.
      c. Difference between this minimal content and actual content as

indication of degree of divorce of any culture from factors of necessity.
    d. Bulk of content of all cultures not a response to necessity.
    e. Culture as an end in itself.
9. The organization of culture.
    a. The limiting factors.
        Needs which must be met.
        Environment.
    b. The dynamic factors.
        Foci of interest.
        Evaluations.
    c. Divergent patterns of culture.

FOR STUDY AND READING

SUGGESTED
  Benedict, "Patterns of Culture."
  Dixon, "The Building of Cultures."
  Kroeber, "Anthropology."
  Linton, "The Study of Man," Ch. XVI-XXV.

MUSEUM ASSIGNMENTS AND OUTDOOR OBSERVATION

A. The Hayden Planetarium

Members of the class are expected individually to attend the Planetarium during two different months during the fall. They should not only attend the lectures in the Zeiss Planetarium but they should also study the Copernican Planetarium and all other exhibits in this building.

B. Evening Field Observations of the Sky

Those who have not been able during the summer to prepare an original sky map will devote a couple of evenings to a study of the sky with the students' astrolabe. Plot on a blank sky map all of the brighter stars observable on these occasions, ignoring their actual names. Then compare with published star chart.

C. American Museum of Natural History

Members of the class are expected individually to visit the American Museum of Natural History, devoting several hours to an examination of the following halls. Study instructions will be distributed:

1. The Darwin Hall. Classification and evolution of the animal kingdom.
2. Hall of Geology and Invertebrate Paleontology. Historical Geology; models; caves; a great copper mine.
3. Hall of Reptiles. Great Dinosaur exhibit.
4. Hall of Mammals. Evolution of mammals.
5. Hall of the Age of Man. Pleistocene.
6. Morgan Hall of Minerals. A magnificent and unrivaled exhibit of crystal forms.

D. One-Day Field Trip

A full-day trip by bus about November 1 or later, across New Jersey to High Point, noting the following physiographic provinces:

1. Manhattan prong of New England.
2. The Triassic lowland.
3. The Reading prong of New England.
4. The folded Appalachians.
5. The Appalachian plateau.

This will provide a cross-section of the chief physiographic provinces of the eastern United States.

E. The Department of Geology Exhibits in Schermerhorn Hall

Type minerals, mine and engineering models. fluorescent minerals, economic minerals, drill cores, fossils, physiographic models, maps, and diagrams.

SUMMER ASSIGNMENTS PRECEDING THE COURSE

The purpose of the summer assignments is to encourage the members of this course to become acquainted with the principal constellations and stars and the planets which may be visible during the summer. The work in the fall, which is necessarily limited in time, will thus be rendered more satisfactory. By starting these observations in June it is possible to extend them over a fairly long period of time and this is important in astronomical studies.

1. Using star chart or planisphere identify chief summer constellations and stars.

2. Optional. Make or buy an astrolabe and prepare an original map of the sky.

3. Indicate on star chart area of sky visible June 1, July 1, August 1, and September 1.

4. Note position of visible planets early in the summer and about once a month thereafter.

5. Note attitude of sun at noon above the southern horizon about June 21, July 21, August 21, and September 21.

6. Observation of eclipses or other astronomical events occurring during the summer.

BIBLIOGRAPHY

Many of the books given in the following list are to be found in the General Science Reading Room and most of them are in the library of the Geology Department. The student will do well, during the course of the semester, to browse through these books, handling each one of them and turning their pages to note the contents. Here and there one should pause to read a paragraph or even a chapter which happens to captivate his fancy. Never again probably will the student find so well selected a set of books right at hand with which to divert himself. As each book is examined it can be checked off the list with a comment, perhaps, as to whether it is a book worth consulting again.

GENERAL

    Mason. The Great Design.
    Newman. Nature of the World and Man.
    Planck. Where is Science Going.
    Powell-Fenner-Bruner. A Survey of Science.
    Saidla and Gibbs. Science and the Scientific Mind.

ASTRONOMY

    Russell-Dugan-Stewart. Astronomy. 2 vols. Highly recommended for all students.

    Alter. Introduction to Practical Astronomy.
    Baker. Astronomy.
         When the Stars Come out.
    Barton and Barton. Guide to the Constellations.
    Fath. Elements of Astronomy.
    Jeans. The Mysterious Universe.
         The Universe Around Us.
         Through Space and Time.
    Johnson. Mathematical Geography. Very helpful on such matters as latitude, longitude, time, etc.
    Marmer. The Tide.
    Martin. The Friendly Stars.
    Menzel. Stars and Planets.
    Mitchell. Eclipses of the Sun.
    Nautical Almanac.
    Serviss. Astronomy with an Opera Glass.
    Swan. The Architecture of the Universe.

CLIMATOLOGY AND METEOROLOGY

    Davis. Meteorology.
    Humphreys. The Physics of the Air.

Kendrew. Climate.
          Climates of the Continents.
Milham. Meteorology.
Shaw. Manual of Meteorology.
Ward. Climates of the United States.

GEOLOGY

Rocks and Minerals

Daly. Igneous Rocks and the Depths of the Earth.
      Igneous Rocks and Their Origin.
Dana. Text Book of Mineralogy.
English. Getting Acquainted with Minerals.
Harker. Petrology for Students.
Iddings. Igneous Rocks.
Kemp. Handbook of Rocks.
Leith. World Minerals and World Politics.
      and Mead. Metamorphic Geology.
Lilley. Economic Geology of Mineral Deposits.
Loomis. Field Book of Rocks and Minerals.
Lindgren. Mineral Deposits.
Merrill. Stones for Building and Decoration.
Moses and Parsons. Mineralogy, Crystallography,
                  and Blowpipe Analysis.
Pirsson. Rocks and Rock Minerals.
Ries. Building Stones.
      Economic Geology.
Tarr. Introductory Economic Geology.
Thom. Coal and Oil - the Keys to the Future.
Tyrell. Principles of Petrology.

Physical, Structural, and Stratigraphic Geology.

Bucher. The Deformation of the Earth's Crust.
Collet. The Structure of the Alps.
Daly. Our Mobile Earth.
Geike. Structural and Field Geology.
Grabau. Principles of Stratigraphy.
Heck. Earthquakes.
Leith and Mead. Structural Geology.
Nevin. Structural Geology.
Peach and Horn. Chapters on the Geology of Scotland.
Twenhofel. Sedimentation.
Wegener. The Origin of the Continents and Oceans.
Willis. Geologic Structures.

Textbooks and Popular Books on Geology

Agar, Flint, and Longwell. Geology from Original
                                 Sources.
Bradley. Autobiography of Earth.
Chamberlin and MacClintock. College Geology. 2 vols.
Croneis and Krumbein. Down to Earth.
Emmons, Thiel, Stauffer, and Allison. Textbook of Geology.

Field. Geology Manual. Two parts, physical and historical.
   Historical Geology from the Regional Standpoint.
Grabau. A Comprehensive Geology. 2 vols. Excellent as a reference.
Longwell, Knopf, and Flint. Physical Geology.
Longwell, Knopf, Flint, Schuchert and Dunbar.
   Outlines of Physical and Historical Geology.
Moore. Historical Geology.
Reeds. The Earth.
Schuchert and Dunbar. Historical Geology.
Scott. Introduction to Geology. 2 vols.
Snider. Earth History.
Williams. Geography of Mother Earth.

History of Geology

Geike. Founders of Geology.
Zittel. History of Geology and Paleontology.

Geomorphology

Bowman. Forest Physiography (of the U.S.).
Coleman. Ice Ages Recent and Ancient.
Cotton. Geomorphology of New Zealand. Part I.
   "Systematic Geomorphology."
Davis. Die Erklarende Beschreibung der Landformen.
   Geographical Essays.
   Physical Geography.
   The Coral Reef Problem.
de Martonne. A Shorter Physical Geography.
Fenneman. Physiography of Western United States.
Hobbs. Earth Features and Their Meaning.
Johnson. Battlefields of the World War.
   Shore Processes and Shoreline Development.
   Stream Sculpture on the Atlantic Slope.
   Topography and Strategy in the War.
Loomis. Physiography of the U.S.
Tarr and Martin. College Physiography.
Tarr and von Engeln. New Physical Geography.

Paleontology

Berry. Paleontology
Bradley. Parade of the Living.
Fenton. The World of Fossils.
Knowlton. Plants of the Past.
Lucas. Animals of the Past.
Lull. Evolution of the Elephants and Mastodons.
   Evolution of the Horse Family.
Osborn. Men of the Old Stone Age.
   The Age of Mammals.
Roemer. Vertebrate Paleontology.
Schimer. Introduction to the Study of Fossils.
Scott. Land Mammals of the Western Hemisphere.
Twenhofel and Shrock. Invertebrate Paleontology.
Zittel. Textbook of Paleontology.

Evolution

Baitsell et al. The Evolution of the Earth and Man.
Darwin. Descent of Man.
      Origin of Species.
Lull. Organic Evolution.
Mason. Creation by Evolution.
Osborn. From the Greeks to Darwin.
Scott. The Theory of Evolution.
Thompson. The Gospel of Evolution.
Wallace. The Geographical Distribution of Animals. 2 vols.
Ward. Evolution for John Doe.

GEOGRAPHY

Geography - General and Regional

Blanchard and Crist. Geography of Europe.
Bowman. The New World.
Brooks. Europe Including the British Isles.
Davis. Handbook of Northern France.
Finch and Trewartha. Elements of Geography.
Huntington and Carlson. The Geographic Basis of Society.
Huntington and Cushing. Principles of Human Geography.
Jones. South America.
Lyde. The Continent of Asia.
      The Continent of Europe.
Ogilvie (ed.). Great Britain; Essays in Regional Geography.
Semple. Geography of the Mediterranean Region.
      Influences of Geographic Environment.
Stamp and Beaver. The British Isles.
Valkenburg and Visher. Geography of Europe.
Willis. Living Africa.

ANTHROPOLOGY

Benedict. Patterns of Culture.
Calverton (ed.). The Making of Man.
Dixon. The Building of Culture.
Hooton. Up from the Ape.
Kroseber. Anthropology.
Linton. The Study of Man.
McCurdy. Human Origins. 2 vols.
Osborn. Men of the Old Stone Age.
Quennell. Everyday Life in the Old Stone Age.
Ripley. Races of Europe.
Schmucker. Man's Life on Earth.
Smith. Human History.
Sollas. Ancient Hunters.
Taylor. Environment and Nation.
      Environment and Race.
Van Loon. Story of Mankind.

BIBLIOGRAPHY

JOURNALS

    Amateur Astronomy.
    American Anthropologist.
    American Antiquity.
    American Journal of Science.
    Anthropologie, L'.
    Bulletin of the Geological Society of America.
    Bulletin of the Hayden Planetarium.
    Economic Geology.
    Geographical Review.
    Journal of Geology.
    Natural History.
    Popular Astronomy.

U.S. GEOLOGICAL SURVEY PUBLICATIONS

    Bulletins, Professional Papers, Folios, Monographs.

Bei Fragen zur Produktsicherheit wenden Sie sich bitte an:
If you have any questions regarding product safety,
please contact:

Walter de Gruyter GmbH
Genthiner Straße 13
10785 Berlin
productsafety@degruyterbrill.com